Murder in Chisago County

THE UNSOLVED JOHNSON FAMILY MYSTERY

BRIAN JOHNSON

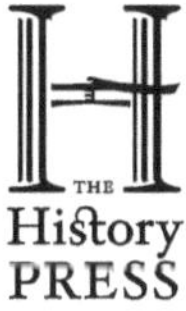

Published by The History Press
Charleston, SC
www.historypress.com

Front cover, top: Albin Johnson as seen in his wanted poster. *Pinkerton Consulting & Investigations, Inc. Bottom, from left:* Harold Johnson, Alvira Johnson and one of the Johnson boys, probably Harold or Kenneth, on a tricycle. *Author's collection.*

Back cover, left: Alvira Johnson holds Harold Johnson. *Author's collection. Right:* Albin Johnson's wanted poster. *Pinkerton Consulting & Investigations, Inc.*

First published 2019

Manufactured in the United States

ISBN 9781467142335

Library of Congress Control Number: 2018963521

In memory of Alvira Lundeen Johnson and her seven children: Harold, Clifford, Kenneth, Dorothy, Bernice, Lester and James Johnson.

Contents

Acknowledgements

This project has literally been decades in the making. Since 1992, when the *ECM Post Review* published my story about the Johnson family mystery, I've flirted with the idea of writing a full-fledged book about the tragedy. At one point, I even started writing a novel based on the events of April 1933. I shelved that idea after realizing that my real skills are in journalism, not fiction. Besides, I could never make up anything as compelling as the true story of the Albin Johnson case.

Life intervened, and the project ended up on the back burner. A big turning point came in about 2007, when North Branch resident Nan Hult contacted me and presented a stack of newspaper articles about the case. She and her friend Dick Lindgren provided a wealth of information. One thing led to another. Before long, I was doing additional research, conducting interviews with old-timers familiar with the story, being interviewed by reporters in East Central Minnesota and speaking in front of a packed house at the North Chisago Historical Society.

I'm indebted to many folks who offered their time, talents and insight. That list includes, among others: Aunt Betty Kollas, Dick Lindgren, Greg Strom, Mark Ruhland, Steve Hansmann, Floyd Pinotti, Jim Carlbom, Barbara Jones, Zac Farber, Venzel Lindholm, Kelly Ann Hokanson, Monte Hanson, Mike Mosedale, Bill Klotz, Derrick Knutson, Ralph Carlson, Ron Hult, Muriel Krantz Kennedy Cash, Al Terry, Mark Johnson, Julie Rowland, Margie Thorp, Gerric Roll, Twyla Ring, Nyle Zikmund, Mark Cohen and David Schultz. Thanks also to the *ECM Post Review* and *Pine City Pioneer* for their coverage of the Johnson tragedy and my book project. I also owe a debt of gratitude to the staff of the North Chisago Historical Society, the Rush City Library, the Harris Lutheran Church and the Minnesota Historical Society.

Many thanks to my wife, Stephanie Johnson, and kids (Victoria, Julia, Lydia and William), who indulged me as I typed away on evenings and weekends and went on road trips up north to do my research. A huge thank-you to Nan Hult for her encouragement, support and research talents. This project would not have happened without her.

Most of all, thanks to my mother, Jeanette Johnson, for sharing her thoughts and memories of the tragedy that claimed the life of her Aunt Alvira and seven cousins.

Introduction

Rush City, Minnesota, may well be the most patriotic small town in Minnesota. The three thousand or so folks who live there don't wait until the Fourth of July to show their national pride. American flags regularly line the main drag, flapping in the wind on most any day of the week, any time of the year.

The town's website boasts that Rush City's first residents included Dakota and Chippewa Natives, who were attracted to the "bountiful hunting and fishing opportunities" there. Logging, fur trapping and farming lured nonnative pioneers to the area.

The city's name is a nod to Rush City's "abundance of bulrushes," or cattails. Incorporated as a village on March 8, 1873, Rush City grew into a mini commercial hub with help from a train, called the Blueberry Special, which brought shoppers in droves across the river from Grantsburg, Wisconsin, the website notes.

For decades, many of those shoppers found lodging or grabbed a hot meal at the iconic Grant House hotel and restaurant. Colonel Russell H. Grant, second cousin to Ulysses S. Grant, put up the original Grant House in 1880. He built the existing three-story brick structure in 1896 after a fire took down the first building. Some folks swear up and down that the place is haunted.

Folks coming into town off Interstate 35 might drive past the Grant House at the corner of Fourth and Bremer. A few blocks past the Grant House, they can hang a left and take a short drive down a country road to the First

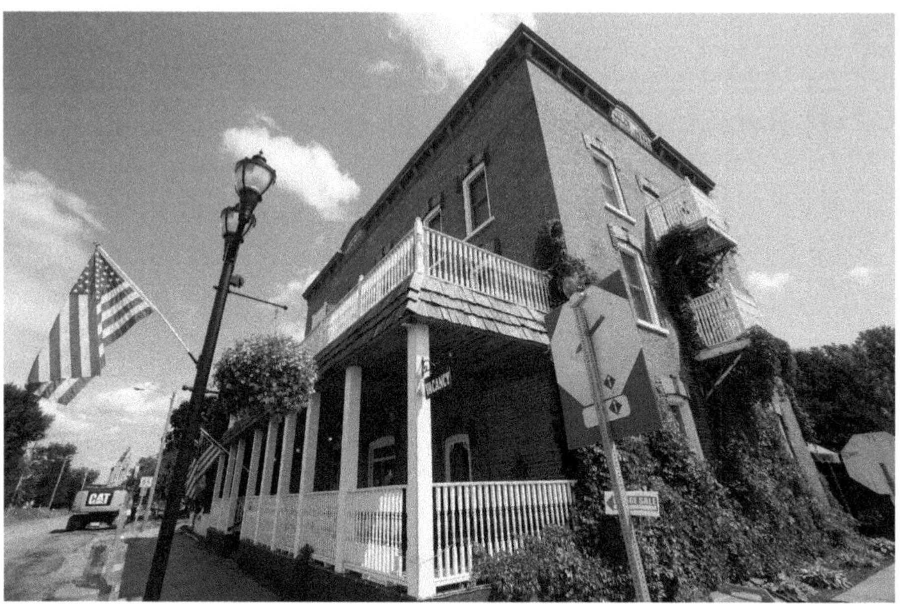

The historic Grant House in Rush City. *Bill Klotz.*

Lutheran Cemetery, final resting place of Alvira, Harold, Clifford, Kenneth, Dorothy, Bernice, Lester and James Johnson. Those family members died as one on April 11, 1933.

The events of April 1933 took place more than thirty years before I was born. But I have always had a spiritual connection to my great-aunt Alvira Lundeen Johnson and her seven children, as well as others who were forever changed by the tragedy that put a chill into that spring day.

It was a Memorial Day ritual for the Johnson family. Every year on that holiday weekend, the whole Johnson gang—Mom, Dad and three kids, of which I was the youngest—would pile into the station wagon and drive about sixty-five miles north of our Minneapolis home to the First Lutheran Cemetery. My great-grandmother Christine Lundeen is laid to rest there, along with other relatives on my mother's side of the family. Next to Lundeen's grave is a nondescript, flat headstone with a simple inscription over the grave of Christine's daughter and grandchildren: "Alvira Lundeen Johnson and her seven children. Died April 10, 1933." (I would find out much later that they actually died on the eleventh. More on that later).

Every year, we would pay our respects at the cemetery. Planting flowers was a big part of that ritual. Pulling with all of our might, we worked the hand-operated pump, located just off a narrow, winding gravel road in

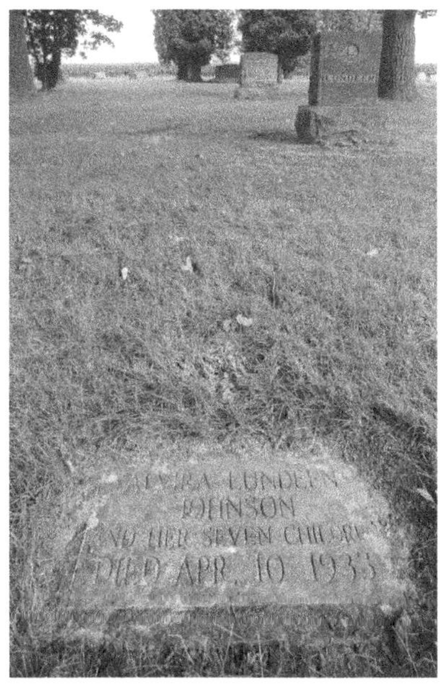

Grave of Alvira Lundeen Johnson and her seven children at the First Lutheran Cemetery, Rush City. *Bill Klotz.*

the rural boneyard, and watched with anticipation as the cool water came gushing out. Then we carefully took the flowers out of their plastic pots, placed them in the ground over the graves and refilled the hole with dirt, patting the dark soil to make sure the flower was firmly supported. We took turns watering the flowers until we were satisfied that they were no longer thirsty.

I remember Great-Grandma Lundeen when she was an old lady. She moaned and groaned a lot and couldn't hear or see very well. Slumped into her wheelchair, with her thinning white hair tied up neatly in a bun, she seemed frail, confused and tired. And yet, she still managed to look content and dignified, despite her advanced age and feeble condition. When I was six, during a visit to Great-Grandma's room at the Green Acres nursing home in North Branch, the elderly lady asked big sister Julie how old she was. "Eleven," Julie replied. "Seven?" Great-Grandma responded. "No," Julie said, louder this time. "*Eleven.*"

On Great-Grandma's 100th birthday, friends and family threw a big party for her at Green Acres. A highlight of the celebration, for the kids anyway, was the oversized birthday cake with a generous coat of white frosting and a "Happy 100th Birthday" greeting spelled out with bright-red icing. One of her relatives had notified the White House of the big event, and Great-Grandma got a congratulatory form letter from President Richard Nixon. A similar greeting arrived from the office of Senator Hubert Humphrey, the Minnesota Democrat who had narrowly lost to Nixon in his bid for the White House.

I'm not sure what Great-Grandma thought of Nixon or the Happy Warrior, as Humphrey was known. But I do know that she had a soft spot for former president John F. Kennedy, whose portrait was displayed on a wall in her dingy, spartanly furnished room at the nursing room. It was close to the picture of Jesus: the one where the Savior is wearing a long, flowing robe and knocking on a door.

To put her long life in a historical perspective, Christine Lundeen was nearly forty-six years old when JFK was born and she was going on ninety-two when Lee Harvey Oswald shot him in Dallas, Texas, with a cheap, mail-order Italian rifle.

She was born in Sweden on Christmas Eve 1871, when Ulysses S. Grant occupied the White House. Every year on or around her birthday, we drove to Green Acres to visit her. Sometimes the weather didn't cooperate, but nothing short of a blizzard with white-out conditions would keep us from making that trek up Interstate 35. Mom always baked a Christmas tree–shaped birthday cake with green frosting for the occasion.

Christine Lundeen died in January 1972, a month after the last Christmas tree cake of her life was consumed. I was eight years old.

Of course, I have no memories of Alvira or her children. And I was too young at the time of Great-Grandma's death to fully grasp the tragic events that she endured well enough to hit the century mark. But those yearly routines at the cemetery and my connection with Great-Grandma made her and those seven children a part of my life. Those rituals made them real to me.

My maternal grandfather, Fred Peterson, was also very real to me. He was married to Freda Lundeen Peterson, one of Alvira's three sisters. I don't recall that Grandpa Peterson ever talked about his mysterious brother-in-law, Albin Johnson. Not that I ever brought the subject up. When I was a kid, I was just happy to watch pro wrestling with him on Saturday night—or, more accurately, to watch *him* watch pro wrestling. Sitting on the edge of his recliner by the fireplace, gesturing in sync with the grapplers, he was a better show than what was on television. And of course, I was thrilled when he gave me some loose change to buy treats at the little market up the steep hill from his place in Oregon. Albin Johnson was the furthest thing from my mind. It wasn't until much, much later that I learned Grandpa Peterson was one of the last people to see Albin Johnson and live to tell about it.

When I was young, I only knew that Alvira and her children had died in a horrific fire. It never occurred to me that it was anything more than that: a tragic, unfortunate accident. As the years went by, I became more and more inquisitive about the family. My curiosity led me to the University of Minnesota's Wilson Library and other research places, where I uncovered old newspaper stories on microfiche that detailed the tragic circumstances, and baffling mystery, surrounding the fire and its aftermath.

By the mid-1980s, I was studying journalism with a minor in Swedish at the University of Minnesota in the Twin Cities. With the zeal of an intrepid

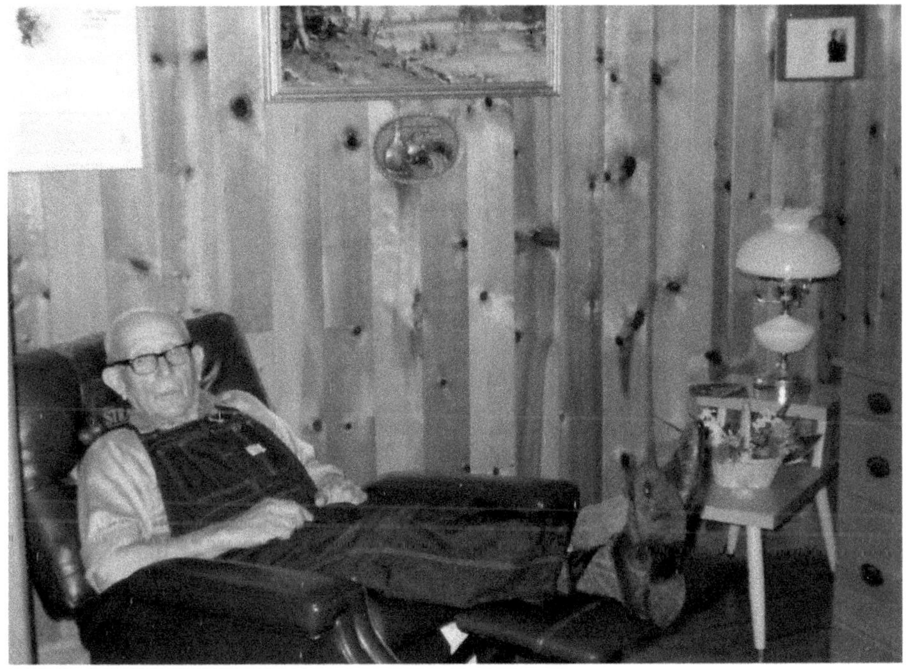

Fred Peterson, brother-in-law of Albin Johnson, at his home in Hood River, Oregon. *Author's collection.*

reporter, I dug up a sheaf of vintage newspaper articles about the Harris fire with help from my brother, Mark. I hoped to eventually use the information as the basis for a newspaper article.

I put the article idea into the back corners of my mind until after I graduated from the university in 1989. In the summer of that year, I accepted a job with American Airlines in Chicago, working as a Swedish interpreter and ticket agent.

It wasn't long before I grew tired of tagging bags and dealing with ornery passengers at one of the world's busiest airports. So in the fall of 1990, I moved back to Minnesota to give writing another try. That's when I returned to the idea of doing a history piece on the Harris fire. I interviewed my mother and others who remembered the fire, including longtime area resident Mae Oscarson and Mae's mother, Elsie Anderson. I also drew heavily on 1930s-era newspaper stories.

I submitted the article in the spring of 1992 to the local paper in Rush City. A few weeks later, I received a copy of the paper in the mail. My article was splashed on the front page, above the fold. A check for ten dollars was enclosed. Not much to write home about, to be sure. But it was the first time I had been paid for writing (not counting the Prince tickets I got as compensation for an internship in the summer of 1988). The article led to my first newspaper job.

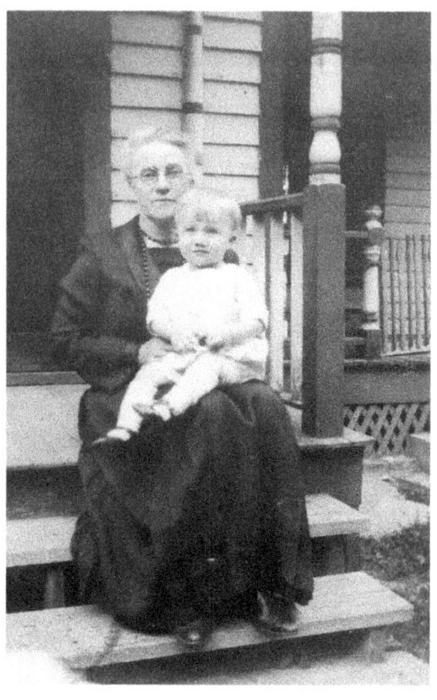

Christine Lundeen and her grandson Harold Johnson. *Author's collection.*

Put another way, the tragedy of the Harris fire touched my life in many respects: childhood trips to the cemetery, a fascination with the story, a published article and the start of a career in writing and editing, for what that's worth.

The fire, of course, touched many people in ways that will never be known.

Christine Lundeen's life experiences included a long and difficult journey to the New World from her native Sweden. She lived through two world wars and the Great Depression and saw the tumultuous 1960s through to the end. She outlived the young president she admired by more than eight years. She raised four children, lost three others while they were still in infancy, buried a four-year-old grandchild and cared for an ailing husband.

And, of course, she lost an adult child and seven young grandchildren to one of the most horrific tragedies in Chisago County history. In my view, it's one of the great untold stories of Minnesota history.

There's a wonderful picture of Great-Grandma Lundeen holding one of the Johnson children. The black-and-white photo portrays a quiet woman of faith. It is clear that her grandchildren were a source of great joy in a life that was often filled with hardship.

Not so long after that photo was taken, she would suffer the mind-boggling experience of losing the child in her arms to a premature death. That same tragedy would claim six other grandchildren and a daughter in the prime of her life. And then there was the missing son-in-law, who was accused of murdering those eight loved ones.

This is the story of that tragedy and its aftermath.

1

The Fire, the Search and a Town Divided

For anyone who has lived through a brutal Minnesota winter, April is a special time of year. After the seemingly interminable months of December, January and February, Old Man Winter finally begins to loosen his vice-like grip on our beloved state. Crocuses and tulips begin to poke through the cold and slushy ground, so recently disguised by a white covering of snow. Sixty with a breeze replaces ten below with a forty-below wind-chill. The days are longer, the nights shorter. Baseball season begins in April, and school starts to wind down. Everyone's favorite team is in contention, and dreams of October glory are within reach.

In April 1933, cheaper phone service was within reach, too. The Minnesota House of Representatives passed a "phone rate" bill, according to the *Rush City Post*. Sponsored by Representative George Johnson of Duluth, the measure limited the cost of one-party business lines to five dollars per month.

That same month, about 1,200 miles southeast of Duluth, the new first lady of the United States took some time to practice her equestrian skills—an outing that didn't turn out well for horse or rider. As the Associated Press reported it, "Mrs. Franklin D. Roosevelt was thrown into a mud puddle in Potomac Park early Thursday when her horse slipped" on a patch of muck and "fell to its knees." "I slid off very gracefully right into the mud," the uninjured first lady quipped, as reported by the Associated Press.

The *Post* also informed readers in April that Chisago County would receive $33,912 for road improvements in 1933. Some of those roads led to the Shadowland Theater in Rush City, where movie buffs could see Douglas Fairbanks in *Mr. Robinson Crusoe* for just ten cents.

Movies weren't the only form of entertainment in town. In 1933, the country took its first steps toward repealing Prohibition, as sales of 3.2 percent beer became legal in nineteen of the forty-eight U.S. states and the District of Columbia. Minnesota was among the states that became wet again.

Many of the drinkers, no doubt, proposed a toast to the new occupant of the White House. While the nation was in the terrible grip of the Great Depression with 25 percent unemployment and despair running rampant, some rays of hope peeked through the dark clouds. A former New York governor named Franklin Delano Roosevelt had just been inaugurated along with his infectious smile, boundless optimism and promises of a New Deal.

But for the Albin Johnson family on the outskirts of the tiny town of Harris, Minnesota, about fifty-five miles north of the Twin Cities, the month of April in the year 1933 brought nothing but raw deals. Even the weather refused to cooperate. Winter had recently made a curtain call in the town of Harris, blanketing the not-so-fertile ground outside the Johnson farmhouse with snow. The snow didn't last long. Even so, spring was put on hold for the time being. But Albin and Alvira Johnson had far greater worries than a delayed spring. While the nation was in the throes of the Depression, the Johnsons and their seven children were in desperate straits. The family of nine was about to be evicted from its farmhouse, a modest, two-story wood-framed structure owned by Albin's father, Emil.

That desperation reached its climax on April 11. During the early morning hours of that grim day, a Tuesday, a raging fire devoured the farmhouse and some of its inhabitants. The remains of Johnson's wife, Alvira, and the family's seven children were found in the ashes of what used to be their home. Albin's body was conspicuously absent. A one-page death certificate, written by an anonymous bureaucrat, sheds little light on the manner of the young mother's death, and it makes no mention of foul play. Issued two weeks after the fire, the certificate says Alvira was a "housewife" and that she was twenty-nine years, four months and twenty-six days old at the time of her death. The certificate states for the record that the cause of death was unknown. "Body burned beyond recognition. Found in burned ruins of home. Origin of fire and cause of death unknown," the document notes in cold bureaucratese.

But that was far from the end of the story. Investigators later concluded that the victims were dead before the flames consumed the house. They

STATE OF MINNESOTA
Division of Vital Statistics
CERTIFICATE OF DEATH

Alvira Lundeen Johnson's death certificate. *State of Minnesota, Division of Vital Statistics.*

speculated that the family may have been murdered, perhaps poisoned, and that the killer subsequently set fire to the home.

(Interestingly, the actual date of death is a mystery in its own right. The fire was discovered in the early morning hours of April 11, and Alvira's death certificate makes it clear that she died on that date. But the date etched into the gravestone reads "April 10." Maybe that was a statement by Alvira's family, who presumably ordered the marker and made the funeral arrangements. If Alvira and the children had been murdered, the Lundeens may have assumed the killings occurred on the tenth and that the home was torched early the next morning.)

The obvious murder suspect was Albin Johnson. If he had died in the fire, investigators surely would have recovered some remains of his six-foot, three-inch 240-pound frame. Authorities meticulously combed through the ashes and debris but failed to find a trace of Albin. Searchers did uncover what was left of a dog that belonged to the family. (A surviving dog, incidentally, would later be the focus of a fire-related conspiracy theory.) If Albin had succumbed to the flames with the rest of his family, it was inconceivable that his bones or bone fragments would not be found in their company.

The mysterious circumstances surrounding the fire set off an all-hands-on-deck manhunt for Albin Johnson, who was eventually indicted in absentia for murder. Wanted posters went up everywhere. Private investigators from the famous Pinkerton agency were on the lookout, offering a fifty-dollar reward for information leading to his capture. The wanted poster said Albin had "worked in Saskatchewan, Canada," and warned that he may attempt to enter the country again. He "has little money," the poster noted, "and may seek food and shelter at relief stations."

Publicity about the terrible blaze and presumed family annihilator spread like a nasty rumor across the country and into Canada, long before the age of social media, talk radio and the Internet. Newspapers as far away as San Antonio, Texas, prominently displayed stories about the fire and the search for Albin. Across the border, Canadian journalists latched onto the story with equal enthusiasm. And in Harris, of course, the story of the missing farmer/suspected killer was on everyone's mind. Some spoke of the tragedy in hushed tones; others got on their soapboxes.

Dick Lindgren, a former local resident who has researched the mystery, first learned about the tragedy in the 1960s, when he inquired about a house for sale in the woods, a half mile from the scene of the fire. Some old-timers with whom he spoke assumed he was talking about the Johnson place and that he wanted to learn more about what had happened. They advised him to leave it alone. At that point, a number of Johnson and Lundeen relatives who remembered the tragedy were still alive, the wounds still fresh. Perhaps the townspeople thought it best not to bring up the touchy subject at all, lest they offend someone.

Cliff Bedell bought the old Johnson property in 1972. No longer a farming operation, the property is strictly a private residence these days. Mr. Bedell is well aware of the grisly backstory associated with the land, but that doesn't seem to bother him. In fact, he has done some investigative reporting of his own. In Bedell's experience, local folks seemed willing to talk about the calamity—but only to a point. "I have talked to old-timers over the years whose first response after finding out where I lived was, 'Oh, that's where the fire was,'" Bedell wrote in a 2007 e-mail to researcher and North Branch resident Nan Hult. "Four men on separate occasions started to tell me they were among the fire site investigators. And no more than a minute of conversation would take place and they all turned and walked away....I'm sure recounting this event brought back horrific memories and caused great discomfort."

The old-timers, like Bedell, have heard all the rumors. Some have speculated that Alvira was pregnant with her eighth child at the time of the fire (not likely). A more macabre bit of hearsay has it that the mother and children were beheaded and the skulls were left behind in the basement of the home (even less likely).

What about Albin's fate?

On that question, the townsfolk were—and still are—divided. Many were convinced their neighbor did the unthinkable and then made a hasty retreat, perhaps on a midnight train to Canada or on a seaworthy vessel on the icy St. Croix River. Albin's critics, especially the Lundeen relatives on Alvira's side of the family, were certain he was still alive. They resolved to track him down and bring him to justice.

Christine Lundeen, Alvira's mother, spent what little money she had on the search for Albin. She hired the Pinkerton agency to see if it could find the wanted man, dead or alive. Amateur bounty hunters, adventurers and ordinary citizens were urged to be on the lookout for the missing farmer.

Others must surely have whispered that the grieving mother was wasting her time. People in that camp insisted that Albin had perished in the fire with the rest of his family. They were either in denial or they simply refused to believe that their neighbor, friend or family member could commit such a hideous crime.

But most everyone agreed on one thing: the fire, the events leading up to it and the disappearance of Big Albin Johnson was a tragedy of epic proportions. As of this writing, more than eight decades after the mysterious blaze, the Harris fire of 1933 remains a riddle.

This is the story of the victims, the missing man and the people who were near and dear to the Johnson and Lundeen families. It draws on interviews from people who remember the fire, news coverage and musings from Harris-area residents.

It all starts with the mysterious fugitive himself.

2
The Johnson Clan and the Professional Boxer

WHO WAS ALBIN JOHNSON?

Albin Johnson, the person, is almost as much of a mystery as the fire that became associated with his name. Fortunately for him, his last name lends itself to anonymity, which would come in handy in his case.

One thing we know for certain: He was a big man. Carrying his 240 pounds on a husky six-foot, three-inch frame, he was a true heavyweight, much like his brother-in-law Farmer Lodge, a professional boxer who sparred with Jack Dempsey and went toe-to-toe with world champions Jack Johnson and Primo Carnera. Born in the dead of winter on January 5, 1890, in Harris, Albin P. Johnson had light-blue eyes and sandy brown hair, according to his World War I draft card. Young Albin was by no means a scholar, but he got at least some rudimentary education as a child. In 1900, ten-year-old Albin Johnson was in school and was able to read and write, according to a U.S. Census report for that year. Years later, his future brother-in-law Harry Galpin noted in an affidavit that Albin's formal education ended after the sixth grade.

According to the 1920 U.S. Census, Albin Johnson was living at the time with his father and mother—Emil and Cecelia Johnson—and four siblings on a farm in the Harris area. Albin was thirty years old. His occupation is listed as "laborer." Though his mother and father came from Sweden, Albin's native language was English, the census report notes. By all appearances, Emil and Cecelia tried to give Albin and his siblings a Christian upbringing.

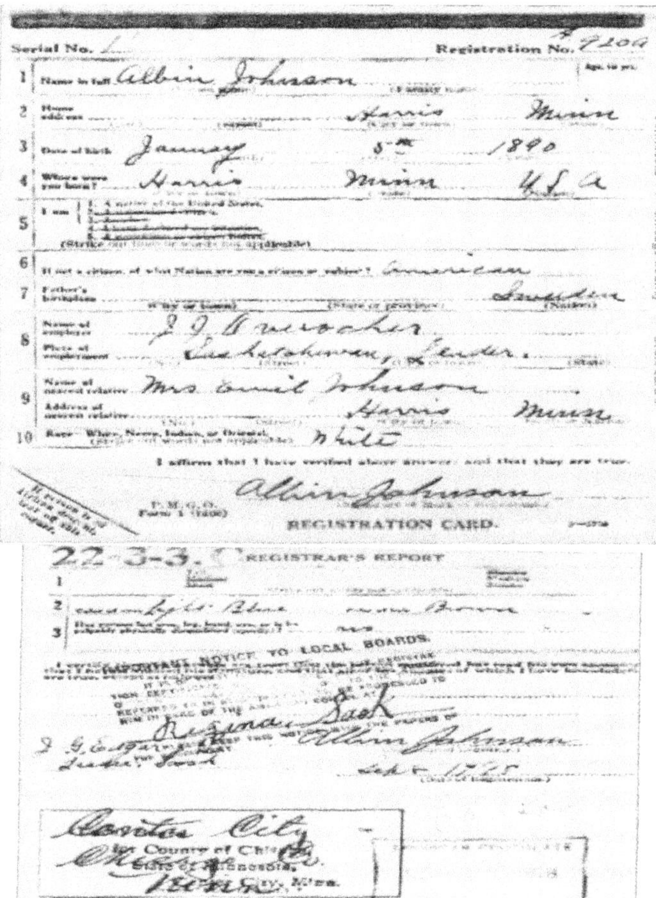

Serial No. 1 Registration No.

1 Name in full Albin Johnson

2 Home address Harris Minn

3 Date of birth 5th 1890

4 Where were you born? Harris Minn U.S.A

6 If not a citizen, of what Nation are you a citizen or subject? American

7 Father's birthplace Sweden

9 Name of nearest relative Mrs Emil Johnson

Address of nearest relative Harris Minn

10 Race White

I affirm that I have verified above answers and that they are true.

Albin Johnson

P.M.G.O. Form 1 (1400)

REGISTRATION CARD.

22-3-3 REGISTRAR'S REPORT

2 Light Blue Brown

IMPORTANT NOTICE TO LOCAL BOARDS

Albin Johnson

Albin Johnson's World War I draft card. *Chisago County, Registrar's Report.*

Albin's youthful face can be seen in a 1904 confirmation class picture from the Harris Lutheran Church. Albin holds down a spot in the middle of the back row, where all the tall kids are told to stand, just a few feet away from his brother Ted. With a thick head of hair parted down the middle in the style of the day, and a stoic, serious expression befitting his Scandinavian heritage, young Albin is quite handsome. He looks like a typical teenage boy in some ways. But his piercing eyes stare back at the photographer with a serious look and a hint of sadness. His female classmates, their hair tied up with bows, occupy the front and middle rows in front of the boys. Staring at the camera with unsmiling faces and flowers pinned to their dresses, they sit primly and hold their Bibles, waiting patiently for the photographer to capture the moment for posterity. Perhaps Albin had a romantic interest in one or more of the

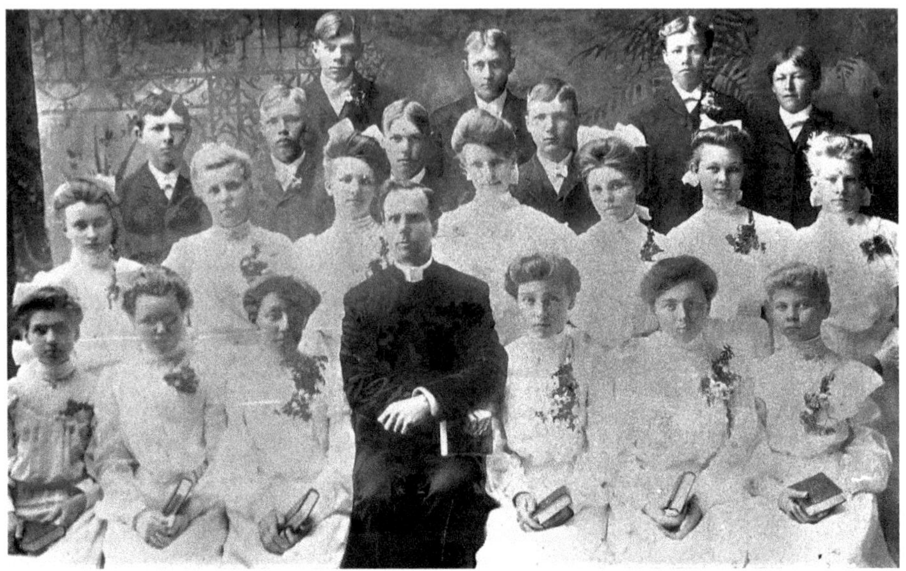

Albin Johnson's confirmation photo. Albin is in the middle of the back row, hair parted in the middle. Ted Johnson is in the back row, far left. *First Lutheran Church, Harris.*

girls—several of whom, incidentally, were named Lundeen. Or maybe he was destined from the start to become involved with another young girl named Alvira Lundeen, who was just a year old when this confirmation photo was taken.

Front and center in the photo is Pastor Linner. With thinning hair and a Bible grasped firmly in his right hand, the good reverend looks to be a man in his early to mid-thirties. He was mature enough to have some experience in the ways of the world, but not too old to have a good rapport with his young charges. Did the pastor see anything in Albin? Did he sense that young Albin may be struggling with some inner demons? Did he try to intervene in some way, perhaps take the young man under his wing? No one knows for sure.

We do know that Albin made his home in Chisago County, which was a magnet for Swedish immigrants in the late 1800s and early 1900s. Author Wilhelm Moberg chose the East Central Minnesota county as the new-world destination for the fictitious Swedish family featured in his popular Emigrants series. In the Moberg books, the main characters are Oskar and Christina Nilsson and their children. Like the Johnsons, they struggled to get by. The winters weren't just cold. They were bitter, intimidating and even deadly in some cases. The work never ended in the farm country, and the rewards for that hard labor were often meager. It was a story of hardship, survival and endurance. The story is set in the 1840s. Not long after, Emil and Cecelia Johnson, parents of Albin Johnson, were among the real-life Swedish immigrants who settled in Chisago County.

Chisago County is tucked in between Pine and Washington Counties on the north and south and is bordered on the east by the great St. Croix River, which separates Minnesota and Wisconsin. The county's name stems from a Chippewa Indian word. Roughly translated, *Ki-Chi-Saga* means "Fair and Lovely Lakes," according to the county's website. Established in 1851 when Minnesota was still a territory, Chisago County attracted its first European settlers around 1680 and later became an important outpost for fur trading and the logging industry.

Swedish immigrants began to settle there in the early 1850s. In *A History of the Swedish-Americans of Minnesota*, published in 1910, author A.E. Strand boasts that Chisago County presented "an agreeable variety" of undulating land that was "covered with hard and soft-wood timber" and "well-watered by lakes and streams." The county's "lake scenery is surpassed in beauty only by some of the lakes in Sweden," the author gushes.

According to Strand, an immigrant named Oscar Roos "could claim the honor of having been, probably, the first Swedish settler not only in Chisago County, but in Minnesota." Roos was born in Västergötland, Sweden, in 1827 and came to America in 1850. He made his way to Minnesota via Rock Island, Illinois. One of his traveling companions, Strand writes, was Lars J. Stark, "who later also settled in this county." The name "Stark," which means "strong" in Swedish, would soon become prominent in the investigation of the Harris fire. But that's getting ahead of the story.

Emil Johnson was born in Älghult, Sweden, on July 1, 1862. According to his obituary in the *Rush City Post Review*, Johnson was "baptized, confirmed, and received his early Christian training" there. Home to only 446 residents as of the 2010 census, Älghult is in the heavily wooded southern Swedish province of Småland. The community has a strong kinship with fire. Like the Swedish town of Orrefors, Älghult is home to a glassblowing factory, where people trained in that craft create beautiful works of art with help from molten glass and fiery furnaces.

Emil Johnson may or may not have been an artist. It's not known exactly what he did in the old country, but it's a safe bet that he struggled to scratch out a living. At minimum, he wasn't hopeful about his future in Sweden. Emil was only twenty years old when he ventured across the ocean to make a new life for himself in the United States. He first stepped foot in the United States in 1882, just seventeen years after the end of the U.S. Civil War and a year after President James A. Garfield was assassinated and replaced in the White House by Chester Arthur.

Emil ended up in Harris, a Chisago County farm town fifty-five miles north of the Twin Cities. He managed to buy some farmland in the new country, and he was in love. In 1888, six years after he arrived in America, he married Cecelia Blomberg, a woman three years his senior. He was twenty-six; she was twenty-nine. Cecelia's family owned a piece of farmland next to Emil's land. After the couple married, they merged the farms.

Cecelia Blomberg Johnson had strong ties to the local church and a desire to serve. Poor health precluded her from serving as much as she would have liked. Cecelia was sixty-four when she died on March 20, 1924. She passed away at the Bethesda Hospital in St. Paul after an illness of several weeks. "After a serious operation, [she] had recovered to the extent that it was thought she could be removed to her home in another week," the *Post Review* obituary states. "Her unexpected death was a distinct shock to members of the immediate family, who were looking for a rapid recovery." Her death notice in the *Rush City Post Review* also states that she was "a member of the Lutheran church at Harris and always took active part in all church affairs whenever her health permitted....She will be greatly missed by all who knew her and loved her for her ever-ready help and kind sympathy in time of need."

Emil outlived Cecelia by twenty-four years. The *Post Review* described Emil at the time of his death as a man "endowed with a strong body and sturdy character" who "contributed much to the upbuilding of this community." Emil was a charter member of the First Lutheran Church in Harris. The Swedish immigrant had a hand in constructing the building, including the majestic bell tower, which continues to produce beautiful chimes that resonate throughout the sleepy town. Some folks who knew the family described them as pillars of the local community. They were well respected and liked, according to the obituary.

Others remember a very different Emil Johnson.

Betty Kollas, a niece of Alvira Johnson, recalls that her grandmother Christine Lundeen never cared much for Emil Johnson. Betty heard rumblings through the years that Emil had an abusive streak. "You got the idea that he was very cruel to his kids," she said. "I had the idea that Grandma and Grandpa Lundeen didn't like him at all."

Floyd Pinotti, a longtime Chisago County resident who acquired the old Ragnar Krantz farm near the Johnson place, is also inclined to question the actions of the elder Johnson. "I don't want to overstate this, but I

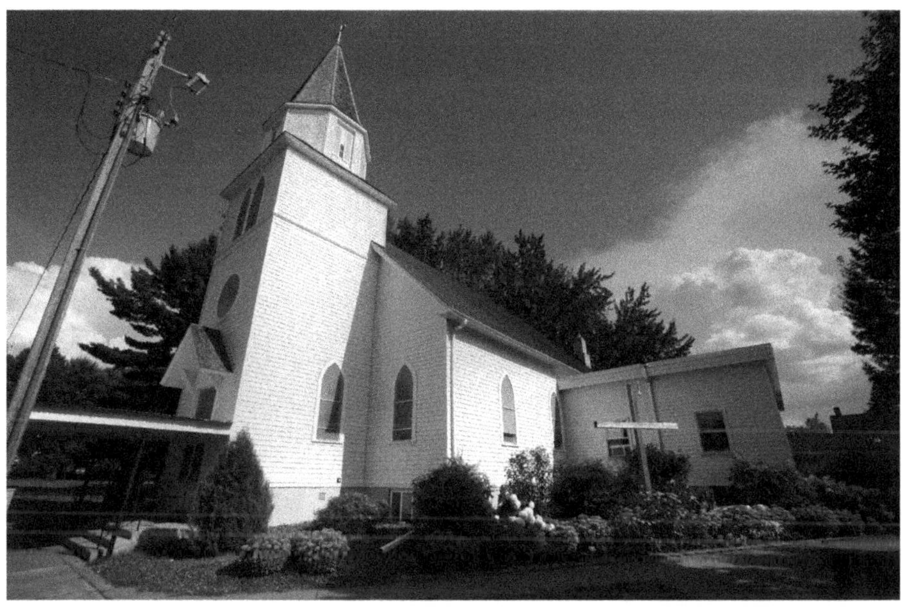

First Lutheran Church, Harris, Minnesota. Emil Johnson helped build the church. *Bill Klotz.*

am a grandfather," Pinotti said in a 2018 interview. "I am a grandfather and a great-grandfather, father of seven and stepfather of two. And if anybody would say I would ever evict one of my children—that would be a catastrophe....During the Depression, when a parent would evict a son from the property who had seven kids, that's a pretty radical thing."

One of Pinotti's stepsons is Steve Hansmann. Though Hansmann never personally knew Emil Johnson, he also got the impression that the old man wasn't exactly a saint. "I always heard that the grandpa, the old man (Emil), was kind of an asshole. He was super righteous, but not a nice person. He wore his faith on his sleeve but he evicted his own grandkids for crying out loud," Hansmann said.

For his part, Dick Lindgren doesn't find it shocking that Emil would evict his own flesh and blood. Money was important to Emil, and the elder Johnson probably counted on the rent he got from the farm, said Lindgren, a longtime Chisago County resident who now lives in Michigan. Numerous reports indicated that Albin was behind on the rent. Albin couldn't make good on his end of the bargain. "I think Emil said, 'Start paying or get out,'" Lindgren speculates. "It doesn't surprise me a bit that Emil kicked them off the place. There was absolutely no money to be had. I could see that happening, where Emil said, 'To hell with you. It's time to let one of the other boys give it a try.'" Given his reputation as a heavy drinker and a no-account rabble-rouser, Albin may have been his own worst enemy when it came to getting his deeply religious father's approval. "Emil was probably a pretty strict disciplinarian. He took his religion pretty seriously. It didn't please him to see his boys acting the way they did," Lindgren said.

Emil Johnson was in poor health for some time near the end of his life. A month before his eighty-sixth birthday, the patriarch of the Johnson family suffered a heart attack. He died shortly thereafter at Rush City Hospital. The real Emil Johnson and the public persona of Emil may well have been two different things. Perhaps there was a dual personality within that body. That could explain, in part, why the Johnson boys had a reputation as roughnecks and hooligans. Could it be that the abused children grew up to be abusive adults?

Long before Emil and Cecelia died, they started a family that eventually included four sons and three daughters. Theodore was the eldest son, followed by Albin, Hjalmer and Henry. In later years, Theodore worked at an auto plant in Flint, Michigan. Hjalmer served with honor in World War II. The daughters were Elsa, Olga and Esther. Esther went on to marry Walter Kenneth "Farmer" Lodge, a professional boxer and wrestler. Standing six-foot-three and weighing 230 pounds, Lodge was one of the more remarkable characters to cross paths with Albin Johnson and company.

Boxing writers of the era often treated Lodge with derision. In the build-up to Lodge's fight with Luis Firpo, 1920s-era sportswriter W.C. Vreeland ridiculed the Minnesotan as a "ham of a fighter" who was willing to "be the chopping block" for the Wild Bull of the Pampas, as Firpo was called. Lodge didn't win any championships, and he never cracked the top ten in the heavyweight rankings. Not especially popular among the fans, he often wore the black hat to the ring in the eyes of the spectators, who cheered lustily for his much smaller, grittier opponents. But he packed a walloping punch and wowed the fans with his sheer size and strength. And no one could accuse Walter "Farmer" Lodge of backing down from a challenge. During a ring career that spanned from 1915 to 1930, a golden era of professional boxing, Lodge took on all comers. He went toe-to-toe with world champions Primo Carnera and Jack Johnson and elite contenders such as Billy Miske, Bill Brennan and Firpo. Though he lost those big fights—and won only thirteen of thirty-five professional bouts in all, according to boxrec.com—he achieved some notoriety as a favorite sparring partner of Jack Dempsey, a megastar of the 1920s and one of the greatest heavyweights ever to lace on a pair of gloves. Outside of the ring, Lodge served his country with honor in the Great War, presumably fighting the Germans with the same resolve and determination he showed in the squared circle. After the war, Lodge found love in the person of Esther Johnson.

Albin Johnson's sister could have done a lot worse when it came to finding a husband. An undated photo of Lodge from his boxing days shows a muscular country boy standing in the orthodox boxing position. He had short-cropped hair and a clean-shaven face that looked like it was carved out of Minnesota granite. He boxed professionally out of St. Paul. It was the sport's heyday, when the legendary Dempsey ruled the ring and big guys like Lodge couldn't resist the urge to show off their toughness in this toughest of sports. Lodge oftentimes boxed Dempsey in exhibition matches. One such match, in Baltimore, was captured on film. The somewhat blurry, black-and-white photo of the two boxers is available in the Library of Congress archives. Lodge is the one in a horizontal position. During his fifteen-year ring career, Lodge fought all over the U.S. and in two foreign countries and scored victories over the likes of Andre Anderson, "Oklahoma Kid" Harvey, "Indian Joe" Cline and Paul Samson Koerner, according to boxrec.com. Lodge's 1930 match with future world champ Primo Carnera, another clumsy giant of a man, didn't go as well. The Ambling Alp dispensed of Lodge in the second round. It was Lodge's last professional bout. The Associated Press reported that Carnera "pushed Farmer Lodge, ponderous Minnesotan, about the ring like a child and then knocked him out in the second round tonight." The ponderous Minnesotan's ring activities weren't limited to boxing. When he wasn't plowing fields or taking punches on the chin, Lodge tried his hand as a professional wrestler. In February 1934, the *Minneapolis Star* reported that Lodge started out as a grappler, turned to boxing and then gave wrestling another shot after "too many knockouts convinced him that the pugilistic game wasn't what it was cut out to be."

Farmer Lodge died on August 5, 1941, when he was only forty-seven years old. Esther Johnson Lodge preceded him in death eleven years earlier. According to his death notice in the *North Branch Review*, Lodge broke his ribs after a fall from a grain stack on his North Branch farm. He later developed pneumonia and went down for his final ten count at the Veterans Hospital in Minneapolis.

Though there's no record that Albin Johnson ever boxed professionally, Betty Kollas, a niece of Alvira, has a hunch he might have tried his luck in the ring to make some extra cash. "That was one way a man could earn a few dollars, was to go into the ring with someone," Betty said. "Albin was desperate for money. He would probably go into the ring. His personality was, 'I could knock anybody out.'"

Indeed, all of the Johnson brothers were arguably every bit as rugged as their pugilistic brother-in-law. The boys were reputed to be rough and tough, and they were big lads, one and all. Some folks have whispered that the Johnson brothers not named Albin may have had something to do with the tragic fire. More on that theory later.

But one thing's for sure: the Johnsons were big and powerful, and you were wise not to cross them. They didn't mind taking a drink, often to excess, and it was rumored that they ran a moonshine operation in Wisconsin. Supposedly, Henry and Ted took turns getting drunk on Friday nights at a local bar. Just off Highway 61, about halfway between Harris and Rush City, folks still drive by a small, unremarkable structure that was frequented by the Johnson brothers. As the story goes, Albin and company walked five miles there to quench their thirst with alcoholic beverages—illicit servings, no doubt, because this was during Prohibition. It took some serious shoe leather to get there. And the return trip must have been a greater adventure as the boys staggered back home.

One of the brothers, in particular, is well-remembered by old-timers in the Harris–Rush City area. The name imprinted on his birth certificate is Henry Arthur Constant Johnson, but most folks in the area remember him simply as "Hank" or "Big Hank." Born on May 26, 1894, in Sunrise Township, Big Hank Johnson was "baptized and confirmed at First Lutheran Church in Harris," according to his death notice. He married his wife, Mary, in 1938. Nine years later, Hank and Mary made their home on the farm place that had belonged to his father, Emil. It was the same land that Albin had unsuccessfully attempted to farm and was forced to vacate just before the fire. Henry Johnson in particular was described as an irascible character. Some said he didn't want any discourse with his neighbors. Others said he would do anything to stay on good terms with those who lived nearby, going as far as to put a bullet into his own dog to keep a neighbor happy.

Mark Ruhland, a Twin Cities resident, has a connection to the Johnson side of the family. His grandmother Mary Johnson was married to two of the Johnson brothers: first Ted, and then Henry. (It must have been quite the small-town scandal in the 1930s for her to leave one brother and then hook up with the other. It's not clear exactly how the breakup played out.) Mark Ruhland said Mary Johnson was "not a logical thinker." She had left

Ted off and on while Mark's mother, Audrey, was growing up. Audrey was still in school when she finally ditched Ted for good and fell into the arms of Big Hank.

By 1933, Emil had retired and moved to town. He turned the farm over to Albin. After the fire, the property sat vacant for a while, and then Big Hank Johnson took over the place in the late 1940s. He built a small house of his own there and settled in. "We remember when we went up there, when we were in grade school, they had a real tiny little house," Ruhland recalled. "The house that my grandma and Hank lived in was a teeny little one-bedroom house. And then around the outside of the house there was kind of a low brick foundation. We remember playing on it....My dad never told us what that was, but that was the foundation for the original house."

Ruhland remembers both Ted and Henry in their later years. When Ruhland was a youngster, his family would head north from suburban Minneapolis to visit Big Hank and Mary once or twice a year. "I lived in Shakopee at that time. That was an eighty-mile drive up there. So we would go a couple of times a year. And we would also see Ted. He lived in Henderson," Ruhland said.

Ted was a hardworking fellow who had a good job in his later years at a car plant in Flint, Michigan. But he wasn't much of a conversationalist, at least not with the children. "Ted, he was a good-looking guy, tall and I think he had good mechanical skills, woodworking skills, stuff like that, that he learned over the years," Ruhland recalled. "But he didn't spend any time talking to us kids. He talked to my folks. We were just sort of cut loose and told to go outside or something. I don't remember ever having a one-to-one conversation with him."

Henry was different. He was kind of "coarse," Ruhland recalled, but he played the guitar and kept the kids amused by doing all sorts of funny things, like wiggling his ears. "We called him Hank," Ruhland said. "Hank was more of a friendly sort. He would do tricks and tell us jokes and entertain us a little bit. Let us ride on the horses. He would take us around the farm a little bit. So we liked Hank."

That likable version of Hank was getting on in years. He was also sober. His demeanor, apparently, could turn ugly after a few too many alcoholic beverages. "I could see that if you poured eight beers into him, he could have been a nasty drunk, too," Ruhland said.

One of the old farmers in town who was all-too-familiar with the rough-and-tumble Johnson brothers was Victor Ramberg, who had 180 acres at the bottom of the hill from the Johnson place. Like any other town, Harris had its share of colorful characters, and Ramberg certainly fit that bill. When Ramberg was well into his eighties, he inquired about purchasing some wooden posts for a fence. The fence would be exposed to the elements, so Ramberg was advised that he needed weather-treated wood to make it last. Ramberg responded, "I'm too old for that," one old-timer recalled. "I don't worry about treated fence posts at my age."

Memories of Vic Ramberg put a smile on the face of another Chisago County native, Steve Hansmann. "He was a good guy—a short little Swede, with huge hands, blue eyes like Paul Newman," recalled Hansmann, the neighbor who grew up near the old Johnson place. "Victor would visit. He gave us our first farm dog. A feral bitch he had shot was killing his chickens, and he gave us one of the pups. And he liked us. He would come out and visit," said Hansmann, a convivial, husky man with a thick white goatee.

A World War I veteran, Ramberg took a machine gun slug in the shoulder when the bullets started flying on the battlefield at Ardennes, France. Years later, he would complain to anyone who would listen, including Hansmann, about his war wounds and the incompetent French surgeon who botched an attempt to put the pieces of his flesh back together. Hansmann was about thirteen when Ramberg pulled him over and told him all about it. "He used to tell me, he said, 'Stevie'—he had kind of a Swedish accent—he said, 'Come over and listen to this.' He moved his arm like this, and you could hear it creak," Hansmann said, making a flapping motion with his left arm. "'You know what that is, Stevie? That's the goddamned French surgeons who wired me up....Those goddamned Frenchies wired me up with some baling wire.' And it sounded like a rusty old machine."

"He was a nice old man. I really liked him," Hansmann added.

Ramberg had a complicated relationship with the Johnson brothers. He must have been friendly enough with Big Hank. After all, he worked the handles as a pallbearer at Henry's funeral in 1967. But he also appeared to be taken aback by Big Hank's temper. Having served in the Great War, Vic Ramberg was no coward. But perhaps the German soldiers were less intimidating than the big farmer from Harris.

Hansmann heard about a particularly ugly incident involving Johnson. As the story goes, a threshing crew came by, and "Henry got mad at somebody and just choked him out. The guy's face was purple. It took four guys to pull him off....I don't even think he was drunk. They were working hard, those threshing crews."

Victor may not have been choked out by Hank. But Victor "was afraid of him. I even sensed that," Hansmann said. "My dad said a couple of times...Big Hank would come over and Victor didn't even want to be in the same room. He [Big Hank] had a reputation. Maybe that's why they were so isolated, too."

But like Ruhland, Hansmann recognized that Big Hank Johnson had a softer side, especially when it came to children. "He was nice to us, and I didn't realize this, because my sister was only four when I left home, but she loved him," Hansmann said. "He would come by, and she would come running up and he would pick her up. He had two sides to him."

Folks new to the area must have had a hard time keeping their Hank Johnsons straight. Besides Albin's brother, there were multiple Hank Johnsons in the immediate vicinity, including Booze or Boozer Hank, Indian Hank and Long Hank or Lanky Hank. Booze Hank was known to go on a bender for a couple of weeks straight before deciding to sober up for a while. One of the Hank Johnsons unrelated to Albin was married to a woman named Jenny, who had a twin brother named Fritz Carlbom. (This is according to longtime Harris-area resident Jim Carlbom, in a summer 2017 discussion with the author.)

Fritz's son, Jim, knew Big Hank—Albin's brother—quite well. As a young man, years after the fire, Jim Carlbom worked for Big Hank, picking corn and doing other farm chores on the property once occupied by Albin and his family. Wearing a baseball cap that said something to the effect that Ford trucks are great and Chevys are junk, Jim Carlbom reminisced about Big Hank and company over a plate full of pancakes on a Wednesday morning at the Kaffe Stuga, a popular diner on the main drag of Harris. For the most part, he remembers Big Hank as a good fellow. "We used to go [to Hank and Mary's place] and watch rasslin' over there. Hank and Mary had a nice little place there. There wasn't a nicer guy than Hank. Oh, my God," Jim said, shaking his head and smiling at the memory.

That's not to say Hank was a pushover. On the contrary, Big Hank wasn't the kind of guy you wanted to pick a fight with, unless you could scrap like Farmer Lodge's buddy Jack Dempsey. When the Johnson boys entered a drinking establishment, it was a bit like a scene out of the Old West, where the local tough guys push through the swinging doors and the dude at the piano suddenly stops playing. "You know, they were rough and tough guys," Carlbom said. "When they walked in the bar, everything went quiet....It did, because if [the other patrons] wanted their ass handed to them, they would hand it to them."

Albin and his brothers worked for a time in the logging camps of Canada. It was physically demanding work, and they found themselves around some rough characters. Perhaps that's where they learned to be scrappy in their own right. Lumbermen worked hard for the few bucks a day they earned in the World War I era. As *The Canadian Encyclopedia* puts it, the working conditions were quite dangerous and fights often broke out among the men, who lived together in tight quarters when they weren't felling trees. "Considering the lumberjacks' cloistered living arrangements and exacting working conditions, it is little wonder they stirred up havoc each spring upon their return to civilization. Stories of them unleashing their pent-up need to brawl, drink and carouse in places like Bytown (now Ottawa) are legendary."

The historical record doesn't indicate how much brawling, drinking and carousing Albin did in his lumberjacking days. But there's reason to believe the man from Harris and his brothers held their own when the going got tough. "They were so damn strong. They were fighters," Carlbom said. "I mean, if somebody wanted to fight—they lived a rough life in the logging camps up there. You took care of yourself. They had a lot of strength there. Hank used to show me how strong he was. Right up to the end he was splitting wood."

Carlbom tells a story about a group of neighbors who, against their better judgment, risked having their asses handed to them by Big Hank—and failed spectacularly. As the story goes, the fellows got liquored up to the point they were brave enough to challenge Hank to a fight. One of the drunks ran halfway across a field toward Big Hank's place before the booze got the best of him and he passed out mid-stride. "He woke up about two hundred, three hundred feet away from the house. He got so goddamned scared he shit his pants and ran all the way home," Carlbom said with a laugh. If officers of the law had managed to track Albin down, perhaps they would have reacted in a similar manner, minus the part about soiling themselves. "I'm sure if the sheriff would have seen him he would have run the other way, too," Carlbom said.

The Johnson brothers weren't the only tough guys in the Harris area. A number of years later, another brawny fellow in town put his muscles to good use to help a cow that had fallen and couldn't get back up. No doubt, the story has been embellished over the years. But as legend has it, the strongman lifted the cow with his bare hands while his friends retrieved a sling.

That same gentleman did a different kind of heavy lifting in a local bar after some other patrons unwisely started to give him and his friends a hard time. "There were six guys in there that started to give them hell, trying to start a fight. They always pick on the big guy. He walked up right between them and patted them on the back so hard he pushed them right down the bar [and said], 'How you guys doing?'" Carlbom said. "There was no more fight in them."

Carlbom recalled that Hank had a strong sense of honor, as well as a reputation as one of the toughest characters in town. After being told that his dog had made a meal of some of the neighbor's chickens, Hank took decisive action. "Hank went outside with him and said, 'Is that the dog, there?' Yup. Hank reached up in the corncrib there, pulled out his gun and...Bam! He says, 'We don't want trouble with the neighbors.' And he loved his dog," Carlbom said.

While at least one of the Johnson boys was willing to kill his own dog to stay on good terms with a neighbor, Albin wasn't capable of taking out his wife and seven children. At least that's what Fritz Carlbom believed. "They said that Hank's brother killed them and everything else. But I don't know. My dad never believed that. He never believed that. He knew them real well," Jim said.

Big Hank died on October 31, Halloween, in 1967. He was seventy-three years old. He's buried alongside his wife, Mary, at Rush City's First Lutheran Cemetery. The modest graves are just a few steps from the final resting place of Alvira Lundeen Johnson and her seven children.

No one will ever know what sorts of secrets Big Hank may have taken to his grave. But Jim Carlbom has an idea of what Hank would say about the Harris fire if the big Swede could rise out of the cold ground for a day.

Hank Johnson always claimed that the searchers didn't look hard enough for Albin's remains after the house burned to the ground, Carlbom said. Furthermore, Big Hank tried to make the case that Albin was in the house when the flames consumed the wooden structure. Hank theorized that the farm dwelling went up in flames so quickly, and the heat was so intense, that there wasn't much left of the bodies afterward. "[Big Hank] said, 'That old house was all built out of white oak, and it just about cremated him.' That's what he thought. It burned so hard," Carlbom said.

What do folks remember about Albin? Because he has been gone for so long, there aren't all that many surviving stories about him, other than some wild urban legends. More on that later. But according to some folks who were around when he was still alive, he had a reputation as a somber, surly man who seldom spoke and was capable of literally taking candy from children. That description seems to match up with the unsmiling, glowering mug portrayed on his wanted poster. "He was morose all the time," said

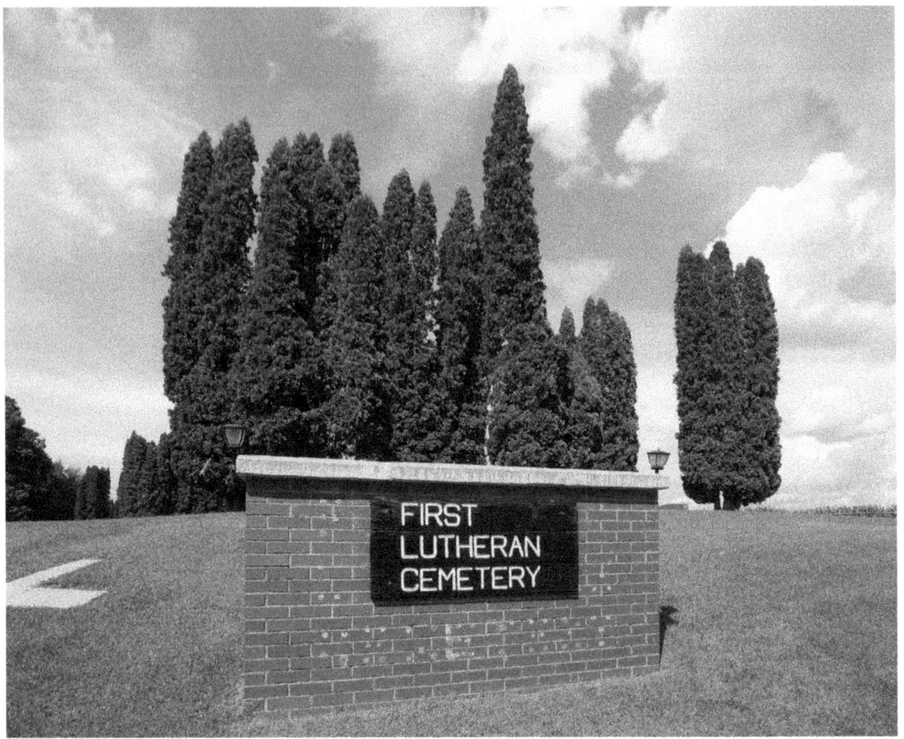

Entrance to the First Lutheran Cemetery, Rush City, Minnesota. *Bill Klotz.*

Jeanette Johnson, Alvira's niece. "Never looked happy, never had a smile on his face." Elsie Anderson, who lived near the Johnson family at the time of the fire, described him in a 1992 interview as "a mean man." Elsie's daughter, Mae Oscarson, remembered him as an introverted character who "never said a word."

Jeanette Johnson heard stories of folks who felt sorry for Albin's family. At times, the shopkeepers would send candy home with Albin to give to the kids. But by the time he got home, the treats were gone. "He would eat the candy and throw the bag away before he got home," she said. "Evidently somebody had called Alvira and asked them if they got the candy. 'No,' she said. 'They never got any candy.'"

It's not clear what Alvira Lundeen saw in Albin, who was fourteen years her senior. "I never heard about why she was attracted to him," Jeanette Johnson said. But the Lundeen and Johnson families lived close to each other in the Harris area. Schoolhouse dances, with a backdrop of music from fiddlers and accordion players, brought people together to have a good time. The atmosphere was festive. The homespun music was comfort food for the ears. Perhaps Alvira and Albin got better acquainted at one of those social functions.

"When we were kids, all the social activities took place in the schoolhouse," Jeanette Johnson recalled. "My sister [Betty], she was just a tiny tot. But she knew this [Swedish folk song] 'Nikolina.' She stood on this stool there and sang 'Nikolina.' And then, those that had instruments could play. My father [Fred Peterson] had a little concertina and somebody else had a violin, and they cleared the chairs aside and had a dance on the floor. That was the center of social activity."

For the Johnson brothers, it's a good bet that the booze flowed freely at those social gatherings. But their dad was more straight-laced. Emil Johnson was considered a pillar of the community. Perhaps Alvira saw some good there. Maybe she thought Albin would take care of her.

Alvira Lundeen was united in marriage with Albin Johnson on December 16, 1922, according to county records. She was nineteen years, one month and one day old when she became Mrs. Albin Johnson. Albin was a few weeks shy of his thirty-third birthday. Christine Lundeen, Alvira's mother, didn't approve of the marriage. Maybe that was because Albin was fourteen years older than Alvira, the baby of the Lundeen family. Or perhaps she had a sense that Albin was capable of being violent. "Grandma had no time for him. She was very disappointed when Alvira married him. She just did not like him," Jeanette Johnson said. "And that's what mother [Freda Peterson, Alvira's sister] said, too....Neighbors around there said that he... never smiled, never had a kind word to say to anybody."

But not everybody saw it that way. Others described Albin as a hardworking man who was devoted to his family. He may have been down on his luck and a bit surly, but he was a good soul at heart, his defenders insisted. A reporter for the *Chisago County Press* was willing to give Albin the benefit of the doubt. In April 1933, the *Press* reported that Albin was "in his usual frame of mind" before the fire, "working hard for his family and taking an interest in his occupation."

At least one neighbor, Caroll Ramberg, was among those who took a kindlier view of Albin. Ramberg apparently told another neighbor, Floyd Pinotti, that he had proof Albin was innocent. As Ramberg understood it, Albin was sitting in a bar called the Old Wagon Wheel at a time that "would have made it impossible for him to commit the murders," Pinotti said in an interview. "He was convinced that Albin didn't commit the murders."

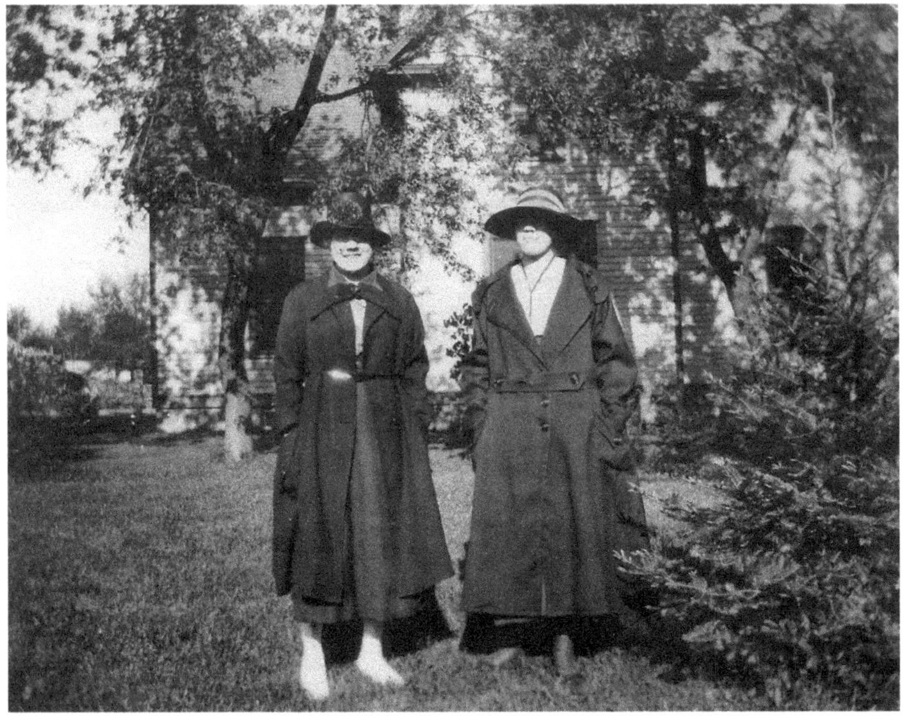

Freda Lundeen Peterson, *left,* and her sister Alvira Lundeen Johnson. Freda had a bad feeling about Albin Johnson. *Author's collection.*

Albin's brother-in-law, Harry Galpin, steadfastly defended the embattled man, going as far as to write a rambling, notarized affidavit that accuses law enforcement and public officials of covering up what really happened on April 11, 1933. Galpin claimed the authorities railroaded an innocent man. As far as Galpin was concerned, Johnson had died in the fire. The investigators, he claimed, either flat-out lied about key facts in the case, such as the conclusion that Johnson's body wasn't found, or they had some other nefarious motive for hiding "the truth" about the incident.

Two years before the fire, Albin's name appeared in the local paper in connection with another incident: the botched robbery of the State Bank of Harris. But in this case he was just an innocent bystander. Albin was a customer in the bank at the time. The cashier was G.J. Stolberg, a relative of the judge who would later play a prominent role in the Albin Johnson case. As the *Brainerd Daily Dispatch* reported it, Bernard Blackfelner wounded cashier Stolberg when the bank employee "refused to obey the command to 'stick 'em up.'" As Stolberg "made a dash for a rear room of the bank to call authorities," the robber shot Stolberg in the arm, the newspaper reported.

The chase was on, and it soon got serious, as the authorities fired on the fleeing bandit in the towns of Wyoming and Forest Lake. After his car broke down in Forest Lake, Blackfelner was forced to "flee on foot for the nearby woods," where he was apprehended by two "garage men," the story continues. In the end, Blackfelner portrayed himself as a lovelorn young man who just wanted some cash to entertain the lady friends in his life. "Girls would not have anything to do with a fellow who doesn't have a car and plenty of cash," the robber whined, as quoted in the *Daily Dispatch*. "I had only 25 cents when I went to the Harris Bank this morning....I wasn't drunk. I just wanted the money to show a couple of girl friends a good time."

Albin Johnson probably wasn't concerned about entertaining any lady friends. But, like Blackfelner, he didn't have much in his bank account. Like so many others of his era, Johnson struggled to support his growing young family. The Johnsons lived in a simple farmhouse, which lacked modern amenities that folks take for granted today.

Betty Kollas, Alvira's niece, was only four when the house went up in flames, but she remembers visiting the site years later with her mother and father. She was ten or eleven at the time. Perhaps it was just a child's imagination running wild, but she recalls that the place gave her the creeps. Young Betty couldn't wait to get out of there. "I got this bad feeling. I was just freezing to death. I told Mom, 'I want to go home.' I had this horrible feeling. We didn't waste any time up there. We took off," she recalled, still shuddering at the memory more than eight decades later. Previously, when the Johnson family was still alive, she remembers at least one visit to the place. "I was pretty young. And I do remember being over there once," Kollas said. "They kept water in big barrels in the house. And somebody lifted me up and showed me that it was water in these barrels. I just panicked. I remember that. I don't think they had a pump close to the house....We never kept any water in barrels."

Lack of indoor plumbing was the least of the family's problems. It was 1933, and the Great Depression was in full swing. Maybe that accounts for Albin Johnson's angry disposition.

American farmers were used to hard times by then. During the 1920s and into the 1930s, a combination of rising debt, falling prices and lower real estate values proved to be a triple whammy for farmers. In Minnesota, from 1926 to 1932, more than 1,400 farms with roughly 258,000 combined acres fell into foreclosure, according to the Minnesota Historical Society.

Hitting the pause button on farm life, the family lived in the Twin Cities for a time. Betty Kollas gets the feeling that Albin wasn't enthused about busting sod or raising cattle for a living. "Albin wanted to work away from the farm. He didn't want to do farming," she said.

Unfortunately for the Johnsons, things didn't work out in the big city. It wasn't long before Johnson and his family returned to Harris. Albin apparently made a good faith effort to find gainful employment. His brother-in-law, Matt Scherer, tried without luck to get Albin hired in the Rush City flour mill where Matt worked. Either the mill had no openings or the mill bosses simply weren't interested in adding Albin to the payroll. It was one of many setbacks.

Johnson and his family settled into the farmhouse owned by Emil, his father. Bad luck, a disastrous economy, poor weather, lack of ability as a farmer or some combination of those factors proved to be Johnson's downfall. "Those were tough years; 1933 was the depth of the Depression," said Ruhland, the Twin Cities resident with family ties to the Johnsons. "The weather was bad, too. And that soil up there is not top-grade soil for farming. So I can see why they may have had a hard time of it. A lot of farmers probably were struggling at that time."

Albin's brother Big Hank took over the farm in the 1940s. Henry milked about a dozen cows and perhaps had some chickens and pigs, local researcher Dick Lindgren said.

People in those parts didn't have much in the way of cash crops in those days. Maybe they grew enough corn to feed the animals, and that was about it. In essence, it was subsistence farming. "They didn't have corn that would grow worth a damn that far north," Lindgren said. Kollas said even the best farmers would have struggled to make a go of it on land that wasn't

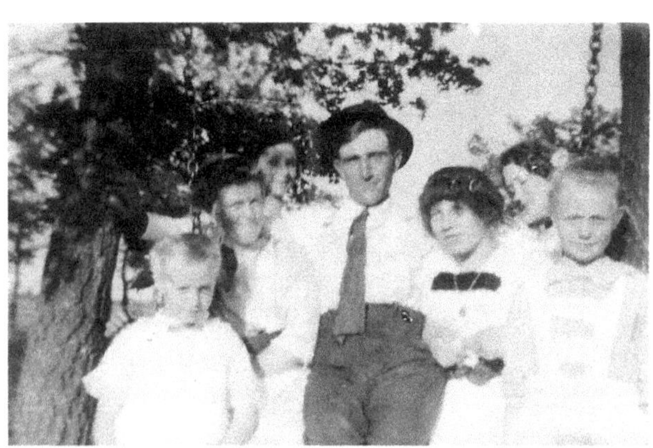

Matt Scherer and Ellen Lundeen Scherer with their children. Matt tried to get Albin a job at a Rush City mill. *Author's collection.*

suitable for crops. The family did have some cattle and a stretch of pasture for the animals. "I couldn't imagine anybody farming on it. Part of it was swampland, part was a hill. You can't grow much on a hill," said Kollas, who owned and operated a ranch in Oregon with her husband, Bud.

Kollas said the county tried in vain to make the land more productive for farmers. "There was this humungous ditch that came right past [the Johnson] place," Betty recalled. "I think the county was putting in a ditch to try and drain the water from that land, so it could be suitable for farming. And then they got this humungous storm that washed the dickens out of everything. It didn't work out very well."

When he was a kid, Hansmann could see the old Johnson property from his yard. Big Hank Johnson was still milking cows by hand there when Steve's family was around in the 1960s. "He milked eight or ten cows by hand. They still had canned milk," Hansmann said. "He had some crops, raised a hog or two to butcher and chickens—just a stereotypical little farmer." But the farm has "never had good crops," Hansmann said, shaking his head at the memory. "Never. There's a lot of clay. The corn would always be stunted, crappy, weedy."

For Albin, things hit bottom in 1933, when Emil Johnson evicted his son, daughter-in-law and grandchildren from the house. The family placed its belongings on a wagon, presumably a horse-drawn vehicle because the Johnsons were too poor to own a truck.

The *St. Paul Dispatch* reported on April 12 that Albin had visited a brother-in-law, Fred Peterson, a day before the fire. Peterson told the newspaper that Albin had said he was "practically set for the moving" and that he, Albin, had "located a farm home near Rush City."

An April 13, 1933 story in the *Chisago County Press* implies that Albin was moving off the farm of his own free will. According to the story, "Mr. Johnson had decided to move off the farm, which was owned by his father, Emil Johnson, and intended to go to another farm

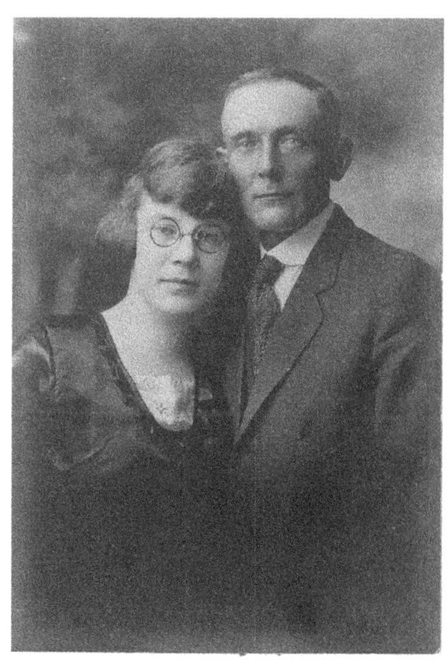

Freda and Fred Peterson. Fred was one of the last people to see Albin before the fire. *Author's collection.*

place near Rush City." But the story contradicts itself in the next sentence, which states that Albin had "received instructions from his father to leave 10 days ago and was packing a load Monday night."

"Can you imagine? It was Albin's father that chased them out," Jeanette Johnson said. "They had no place to go."

As of Monday, April 10, the move appeared to be imminent. The family was, by all accounts, ready to say goodbye to the old farm place. The Johnsons had "beds, furniture, and other household goods in a wagon which stood outside the home and was also burned," the *Chisago County Press* reported. The wagon stood on the property in the early morning hours of April 11, 1933.

That's when the fire was discovered.

3

Flames in the Night, Smoldering Ruins and a Massive Manhunt

The farm place that Albin and Alvira Johnson called home was bordered roughly by Highway 61 to the west, 470th Street to the north, the Government Road to the east and 460th Street to the south. A big swamp and a prominent high ground, known as Chippewa Hill, lie just to the east. To the north is a large ravine, which separated the Albin Johnson place from "Booze" Hank Johnson's residence. Or as the *North Branch Review* described it in 1933, the Johnson farm was "about four miles northeast of Harris, the home being about two miles east of the so-called five-mile curve."

The hundred-acre Krantz farm was immediately to the south of Albin's place on the other side of the narrow, gravelly road called 460th Street. More than eight decades after the fire, cars motoring down that road still kick up dust as they whiz by the open spaces interspersed with stands of trees, grazing cattle and the occasional farmhouse, barn and silo. Ragnar Krantz lived in that neighboring farmhouse with his wife, Anna Marie Carlson Krantz, and their young children.

The Krantz family had deep roots in the community. Ragnar's father, Pedro Johan Krantz, immigrated to the United States from Sweden in 1880. He and Maria bought three hundred acres of Harris-area farmland from the railroad and settled there. "He gave one hundred to my dad, one hundred to my uncle and a hundred to another uncle. We were all three in the same area," said Ragnar and Anna's daughter, Muriel Krantz Kennedy Cash.

Ragnar was an industrious man of many talents and an upstanding community member. When he wasn't tending to his cattle, pigs and chickens, Mr. Krantz was a deacon at the First Lutheran Church in Harris. Pedro Johan helped establish the church in the early 1890s. Ragnar also served as a constable for Sunrise Township, a tiny Chisago County village on the shores of the Sunrise and St. Croix Rivers. "His main job was to keep the peace at the elections. Of course, there were never any big fights at Sunrise," Muriel said with a smile. "But he would go early in the morning and stay all day. It was a big deal for him."

During the early morning hours of April 11, 1933, Krantz was most likely sound asleep, along with twenty-one-month-old Muriel, five-year-old Willard and the rest of his family. Suddenly, flashing lights cut through the darkness. He woke up to discover a strange vehicle on his property and his neighbor's home engulfed in flames. An April 15, 1933 *Minneapolis Star* report shed some light on the circumstances leading up to Krantz's discovery of the flaming house. Fred Trippen, a local farmer, told the United Press that he saw the flames around 3:00 a.m. while driving by the Johnson place on Highway 1, the *Star* reported. Trippen then drove onto the Krantz property, "flashed headlights into the window" of the home and "directed Krantz's attention to the flames," the newspaper added. Krantz told the *Tribune* that he first noticed the fire around 3:30 a.m. His actions were heroic and he acted quickly, but it was too late to help anyone. "After giving the alarm, I drove over to the Johnson place as fast as I could get there, but the house by that time was almost totally destroyed. Only one corner remained standing, and after a short while that crumpled, too," Krantz said as quoted in the April 12, 1933 edition of the *Minneapolis Tribune*. As if the Johnson family didn't have enough bad luck, even the weather conspired against them. In its April 12 report on the conflagration, the *Duluth Herald* observed that "a stiff breeze" fanned the flames, which quickly "wiped out the home."

Krantz offered at least one tantalizing clue that seemed to suggest the presence of a getaway car at the scene. The *Pioneer Press* reported on April 15 that Krantz had seen "automobile tracks leading from the Johnson driveway onto a highway that had been freshly dragged at dusk" on Monday, the day before the fire. But the clue, like Johnson himself, quickly vanished. According to the newspaper, the tracks were "obliterated and it was impossible to trace them" after "many other vehicles were driven in the farm yard." The *Minneapolis Star* story rules out the possibility that Trippen's vehicle could have been responsible for the tracks. After alerting Krantz, Trippen "drove on," which "disposed of the possibility that Trippen's automobile made the telltale tracks," the newspaper reported.

Muriel was too young to remember that dreadful scene. But her now-deceased brother, Willard, took it all in. "My brother was about five. He used to say he remembered the reflection of the flames on the wall, but he was five years old. You don't know," Muriel said.

During the investigation, the sheriff and his cohorts used the Krantz place as a staging area because the family had a telephone. "They were coming in there to use the phone and to have coffee and to get warmed up," said Lindgren, the local researcher, who spoke with Willard Krantz about the fire in the mid-2000s. The younger Krantz remembered the sheriff as "one big, tough-looking hombre. He will always remember those high laced-up boots that the sheriff wore and the big hat," Lindgren said.

Willard Krantz went on to live a fruitful and successful life. After graduating from Rush City High School in the 1940s, he earned degrees at the University of Michigan and the University of Hawaii, where he eventually settled. Willard Krantz was a public information officer at Fort Shafter, Army Command for the Pacific. He and his wife, Junko, had a son and a daughter. During a career with the U.S. Army and Air Force, Ragnar's son served on American military bases in places like Japan, South Korea and Saudi Arabia, a world away from the yawning countryside of East Central Minnesota. He died in 2010 at age eighty-two.

But even after all of those life experiences, Willard Krantz could never shake the eerie feeling of seeing his neighbor's house burn. Decades later, Willard claimed to still have vivid memories of that terrible night. It was almost like he could still see the flames, feel the heat and inhale the suffocating smoke seventy-plus years afterward. He shared some of those dark recollections with Nan Hult, a local resident who has done extensive research on the fire and its aftermath.

"It's something he said he will never forget," Hult said at her home in North Branch. "He remembers waking up and seeing the fire on the wall. He said it looked like blood dripping down....He saw the fire coming off the windows, and it looked like molten lava or something going down the window. On the bedroom wall, he said, it looked like it was flickering."

Jeanette Johnson, Alvira's niece, also remembers the fire, though her recollections aren't quite so vivid or creepy. As of 2018, Jeanette is ninety-seven years old and living in a senior-care facility in Minneapolis. Her third-floor room is small and cozy—about the same size as Christine Lundeen's dwelling at the former poor farm known as Green Acres. But the comparisons end there. While Mrs. Lundeen's room was dull and drab, Jeanette's has all the comforts of home with family photos, a quilt and artwork from her grandchildren on the walls. There's a small table, a bed and an antique dresser topped with a framed photo of her beloved husband, Willard Johnson, looking dapper in his World War II army uniform. A large window overlooks a play area for the building, which also has a childcare center. Jeanette spends much of her time sitting near that window in a comfortable reclining chair. Her mind is still sharp. Though she has a TV for entertainment and a remote control within arm's reach, she prefers to read. A bookshelf in the corner holds some favorite reading material, including selections in Swedish, her first language.

Jeanette was twelve years old when the Albin Johnson farmhouse went up in flames. Though some memories have faded, she recalls her mother's and grandmother's horror as they heard the terrible news. "Ragnar Krantz called right away. He saw the fire. They said the fire lit up the whole area there and he happened to see it. And he called [Christine Lundeen, Alvira's mother] then right away," Jeanette said as she leaned back in her favorite chair.

It was still pitch dark when young Jeanette woke up and heard the commotion. "[My mother was] talking on the phone and crying. It's a morning I will always remember. I can even remember the place where I was sleeping. I was sleeping in the living room instead of my bedroom....I don't know why I was downstairs when I was awakened," she said.

After Christine Lundeen heard the news, she called another daughter, Olga, who was living in Minneapolis. Olga "was shocked. And Bruce, her son, told me about that. When his mother answered the phone, she just fainted from the shock," Jeanette said.

Frank Hanson, the chief of the Rush City Fire Department, rushed to the scene of the fire and arrived within twenty-five minutes, according to the *Rush City Post*. Not bad for a small-town fire operation in the 1930s, especially considering that the dirt roads leading to the farm had most likely turned to mud and slop in those early days of spring.

"You have to remember, there was fresh snow," said Chisago County native Greg Strom, who has researched the tragedy. "There were no snowplows back in 1933." Hanson probably didn't drive a snowplow. But the eldest child of Danish immigrants Peter and Cleria Hanson was about four years into a thirteen-year run as the town's top firefighter, and he wore many hats. Besides serving as fire chief, Hanson was an undertaker, grocer, restaurateur, Chisago County commissioner, Rush City postmaster and volunteer fireman in Rush City for nearly half a century, according to the *Rush City Post*.

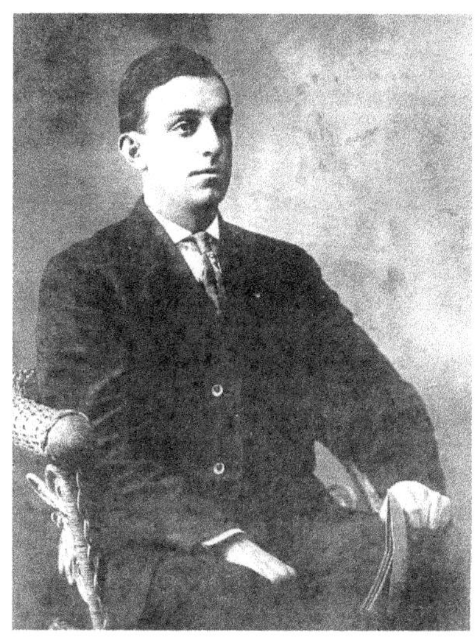

Frank Hanson served as Rush City's fire chief for thirteen years. *North Chisago Historical Society.*

Hanson, who had lived on his own since he was fourteen, had a daughter with his wife of nearly thirty years, the former Henrietta Dorothea Kloock. He resided in the area for nearly fifty years. Hanson "had identified himself with numerous civic enterprises, whereby he had gained for himself a very wide circle of friends," the *Rush City Post Review* reported at the time of his death.

Hanson died on October 26, 1942, less than ten years after the calamitous fire. Though Hanson's passing wasn't as mysterious as that of Alvira and her seven children, it's a story worth telling. Wearing his undertaker hat, Hanson was en route to St. Paul to pick up the remains of a newly deceased gentleman named John M. Kroeger and to return the body to Rush City for burial, according to an October 30, 1942 article in the *Post Review*. The *Post Review* picks up the story from there: "With Henry Hanson at the wheel of his hearse, death's summons came as Frank slept at Henry's side. Death occurred at approximately 5:57 p.m., his age at death being 67 years, 5 months and 19 days."

Before he died riding shotgun in a hearse, Hanson was a go-to guy in the community, a source of advice and wise counsel. As the *Rush City Post Review* reported it, he was "such a willing public servant that it became commonplace for folks of the Rush City community to say, when some problem arose, 'Let's go and talk to Frank Hanson about it.'"

But the problem that rudely awoke Frank Hanson in the wee hours of the morning on April 11, 1933, was too great even for him to handle. Indeed, by the time he arrived on the scene, it was too late to rescue anyone. "At that time, the house was completely demolished, with the exception of one corner, and nothing could be done to save any portion of it," the Rush City newspaper reported.

Hanson wasn't the first to arrive. Four neighbors were already at the site of the disaster, helpless to do much of anything but stare at the terrible scene unfolding before their eyes.

Hanson went into action. As a civic leader and fire chief in a rural area where everybody knew everybody, he undoubtedly was aware that the farmhouse was home to a young family. Local newspapers reported that he searched the outbuildings, surrounding fields and woods for signs of any surviving family members.

At daybreak, Chief Hanson left seven or eight men to watch the premises and alerted the deputy coroner. He was probably happy to get away from the grisly scene. As the fire raged on, "one of the bodies could be seen in the burning embers," the Rush City newspaper reported in graphic detail.

Within the ruins, searchers found the charred remains of Alvira Johnson and her seven children: ten-year-old Harold, nine-year-old Clifford, seven-year-old Kenneth, five-year-old Dorothy, four-year-old Bernice, two-year-old Lester and the baby of the family, four-month-old James.

Albin Johnson, Alvira's forty-three-year-old husband and the father of the children, was nowhere to be found. It was as if the rural farmland had gobbled up the big man.

Naturally, the story was huge news in the local paper. And there was a strong hint of mystery even in that very first article. Under a bold front-page, all-caps headline that read, "Family of Eight Perish in Flames," the Rush City paper reported there was "no trace of" Albin Johnson. The newspaper, in an understatement, described it as "one of the greatest tragedies that has occurred in this vicinity for several years." Other media outlets, from St. Paul to Saskatchewan, from Brainerd to Bismarck, picked up on the news. Some reports, perhaps in their haste to get the story, got a number of facts wrong.

An April 11 Associated Press story assumed that Albin was one of the victims. According to the AP report, "Nine persons, including an entire family of father, mother and seven children, were found burned to death today when fire destroyed their farm home." (Albin's defenders, such as brother-in-law Harry Galpin, would insist that Albin had, indeed, died in the fire. Galpin's theories will be explored later.) The news service reported that the victims were Albin Johnson, "about 46," as well as "Mrs. Albin Johnson" and seven children "from four months to 10 years." (In fact, Albin was 43 at the time of the fire.)

Another AP story, which ran three days after the fire, presented an even more sinister narrative, in which the mother and children perished in the fire and the father was on the lam. Under the headline "Mother and 7 Children Die; Man Missing," the story declared that Chisago County authorities had launched an investigation.

That's where A.O. Stark came into the picture.

Born to early Chisago County pioneers on April 22, 1871, in nearby Fish Lake Township, Albert O. Stark may well have been called "Mr. Harris."

His father, Lars J. Stark, was a traveling companion of Oscar Roos, who was believed to be the first Swedish settler in Chisago County, perhaps all of Minnesota. Along with his wife, Adelaide, and three sons, A.O. Stark lived in Harris most of his life. To this day, the family's imprint is omnipresent. Folks drive down Stark Road and watch kids play Little League baseball on Stark Field. There's even an entire town named after the Stark clan.

Stark was a businessman, a community leader, a licensed mortician and an active member of the Harris Lutheran Church. In a career that stretched from 1893 to 1945, he dabbled in the lumber, banking, hardware and funeral businesses. Stark and two of his brothers owned the Harris Hardware Store, where they also sold lumber and agricultural machinery. The hardware store, a big, white building, doubled as a makeshift morgue and mortuary back in the day. Not content to peddle construction and farm equipment, he also owned twenty shares in the Harris State Bank—an investment that would later raise the hackles and suspicions of Harry Galpin, Albin Johnson's tireless defender and devoted brother-in-law. Stark's April 1961 death notice doesn't list farming as an occupation. But it does mention that Stark was the "last living member of the first graduating class of the School of Agriculture at the University of Minnesota."

In April 1933, Stark was called to serve the community in yet another capacity: deputy coroner of Chisago County. Leonard J. Lund, deputy state fire marshal, assisted. Lund, of Minneapolis, left for Harris about a week after the fire to "take charge of the hunt," according to the April 18, 1933 edition of the *Minneapolis Journal*.

Stark was the first investigator to raise suspicions about Albin. The deputy coroner, who oversaw a search of the ruins, declared just a day after the fire that Albin's body was not in the ashes of the home. "From all indications we have, Johnson was not in the house when it burned," Stark said, as quoted in the *Duluth Herald Tribune*.

Stark met with County Attorney S.B. Wennerberg to discuss next steps. While Stark was sure from the start that Johnson was on the loose, Wennerberg was more cautious—at least in his early statements to the press. "Everything found so far indicates the fire was accidental," he said, as quoted in an April 12 Associated Press story. "If we cannot find Johnson's body by tonight we probably will proceed on the theory he disappeared from home before the fire."

A longtime resident of Center City, S. Bernard Wennerberg was born in 1894 to August and Anna Wennerberg. The younger Wennerberg was a law-and-order guy, and he didn't back down from a challenge. Though he was far from the big gangster hangouts like Chicago and New York, he wasn't afraid to call out mobsters and folks in high places. While former New York governor Thomas Dewey made a name for himself as a gangbuster, a reputation that almost took him to the White House, Wennerberg verbally went after the mob in a December 1933 interview on WCCO radio. As reported by the *Minneapolis Star*, Wennerberg claimed in his interview that an "army" of 400,000 criminals was responsible for 12,000 deaths in 1932. He blasted "men in high office" who have "condoned mob violence," the newspaper noted. Albin's pursuer was about eighty years ahead of his time when he called for "strict supervision over manufacture and sale of firearms" and "arousing a public heretofore content to forget about election times." Wennerberg himself was seen as a rising star in politics. In April 1934, a year after the fire, he was the keynote speaker at the GOP endorsing convention at the Minneapolis Auditorium, where he "bitterly attacked" the rival Farmer Labor Party and declared that the Farmer Laborites had "become drunk with power," the *Minneapolis Tribune* reported at the time.

Years earlier, he attended Gustavus Adolphus College in St. Peter, Minnesota, where his activities included a stint on the school's speech team. A March 1913 article in the *Manitou Messenger* notes that Wennerberg took part in the annual contest of the "Inter-Collegiate Oratorical Association of Minnesota" at Central Presbyterian Church in St. Paul. Other college speakers waxed eloquent on themes such as "American Ideals," the "rather threadbare topic" of "Law Enforcement in the United States" and the "Commercialization of the Press," the *Messenger* reported. In that last topic, speaker Gustave A. Stenerson insisted that the "press of today is the slave of 'Big Business' and to the advertiser. He set forth the office of the press as the molder of public opinion and pleaded for the true use of this great influence." But Wennerberg led his Gusties to victory with a speech titled "The Appeal of Labor." Wennerberg's choice of topics implies a strong interest in fighting on behalf of the little guy, like the humble farmhand who made his living by the sweat of his brow. In his speech, a "strong appeal for justice to the toiler was made," the newspaper reported. "Mr. Wennerberg had a very pleasing appearance and delivery and delivered his message with ease and earnestness."

The speaking skills Wennerberg honed at Gustavus served him well in later years. During the Albin Johnson investigation, he was the name, face and voice of the team to a great extent.

In 1933, Wennerberg's definition of justice was tracking down the alleged family annihilator, Albin Johnson. His quest for justice may well have been earnest, but it certainly wasn't easy.

The investigators confirmed early on that Emil Johnson, Albin's father, owned the farm and that the younger Johnson and his family had packed their belongings and had plans to move. County officials didn't disclose the family's destination or the reason for the move.

Investigators did reveal that the bodies of the mother and one child were found together. Five others were in a separate group, and one was discovered in the basement, presumably landing there after the floor gave way to the flames. They were sleeping on makeshift beds, which is understandable because their regular beds were likely stacked up with other furniture on the wagon outside the house. Presumably, the child who slept with Alvira was James, the baby.

On April 13, the *Chisago County Press* reported that the "party of searchers under the direction of Coroner A.O. Stark of Harris, after having rummaged through the ashes and debris of the burned farm house, extended their hunt to places about the premises with no result," the newspaper reported. "Charred bones, all that remained of the other members of the family, were found and it was learned that the remains of the mother and seven children had all been accounted for. A small lake near the place was searched as was the adjoining timber land for the missing father," the *Press* article continues.

Venzel Lindholm, an old-time Harris-area resident who was five at the time of the fire, recalled over a cup of coffee at the Stuga in 2017 that some of the searchers were armed with an unusual tool. "There were guys… poking into the hay with pitchforks, figuring maybe he had been hiding underneath the hay," Lindholm said between sips of hot coffee.

Lindholm's father, Francis, accurately predicted it would be tough sledding. The elder Lindholm had a sense early on that searchers wouldn't have much luck finding the AWOL farmer, with or without pitchforks. "I remember my dad saying, 'That little investigator, that little short guy chewing on a cigar…he won't ever find him.' But that was my dad," Lindholm said with a verbal shrug.

While the short guy chomping on the stogie and the other searchers combed the countryside in vain for traces of the man with a fifty-dollar price on his head, the wheels were starting to turn on the criminal justice side. The authorities didn't speculate on a motive for the alleged killings. Perhaps they knew more than they were letting on. If they had information, they were keeping it close to the vest. But more details began to emerge at the inquest, which was held in mid-April. At the inquest, a coroner's jury voted an "open verdict" in the case, which meant in essence that the deaths were of "unknown cause"—an important classification that didn't rule out murder.

Hank Johnson, Albin's brother, testified at the inquest that he had loaned twenty dollars to Albin on the Saturday before the fire to use for the first month's rent on a new home on a farm near Rush City, though it was "brought out" that the rent was never paid, the *St. Paul Pioneer Press* reported at the time. Johnson also mentioned that Albin had expressed a desire to return to Canada, where he had worked in 1917. "Henry Johnson had talked about a trip they had taken to Canada a few years ago, and that he [Albin] remarked on the desirability of being there now," the *Pioneer Press* reported.

A Red Cross worker named Mrs. A. Hunnecke testified that she had visited the Albin Johnson home recently and that the family was "not in want," and Ragnar Krantz mentioned those mysterious tire tracks heading from the Johnson place to the highway on the night of the fire, the newspaper reported.

Also at the inquest, Emil Johnson came clean on the fact that he had kicked Albin and family off the farm. The elder Johnson testified that he had "asked his son to vacate for non-payment of rent and that his son had transferred all his personal property to him" on the Saturday before the fire, the *St. Paul Pioneer Press* reported on April 15.

On that same day, a *Minneapolis Tribune* story offered additional details about Emil Johnson's decision to evict the family. Put simply, Albin was quite the deadbeat tenant if Emil's testimony was true. The elder Johnson ordered Albin to "move from the farm for failing to pay rent over four and a half years," the newspaper reported.

Furthermore, the story strongly implies that the borrowed rent money was about to go up in smoke, so to speak. "Albin arranged to rent a Rush City house for $12 a month, and borrowed $20 from two brothers, but failed to pay the rent," the *Star* story continues. "The farmer spent several hours in Rush City Monday [the day before the fire], making three separate purchases of tobacco, totaling eight sacks."

In a cryptic postscript, the newspaper reported that Albin appeared in Harris the day before the fire and asked "what time the next bus went north, but did not stay to meet it. He saw his father and deeded over cattle and farm equipment so the father would assume a joint $800 note."

The *Minneapolis Star* also reported some graphic details about the coroner's jury proceedings. According to the newspaper, "A casket containing the remains of the eight victims was wheeled onto the town hall stage at the inquest. Later, for members of the coroner's jury, the casket was opened on two occasions." After the casket was opened the second time, a "buttermaker" named N.A. Neilson "voiced a belief that the ashes of an adult other than

Mrs. Johnson" might have been discovered, the newspaper reported. Neilson, who assisted with the search for the missing farmer, thus implied that Albin was among the dead. The coroner's jury disagreed. "While Neilson and the jurors crowded around, the casket lid was raised, but a few minutes later the verdict accounting for only Mrs. Johnson and the children was read," the *Star* reported.

In a sense, the inquest raised more questions than answers. Who was driving the vehicle that left tracks on the Johnson property? Where was it headed? And why had Albin expressed a desire to get out of Dodge before the events of April 11, 1933?

Dick Lindgren, the local researcher, recalls what Willard Krantz remembered about the mysterious tracks. "Willard Krantz said that when they woke up and were in the process of reporting the fire, a car drove in the dirt road in front of their place and drove into the Johnson driveway, which I suppose was nothing but a big mud hole that time of year," Lindgren said. He speculates that the driver was spooked by the flurry of activity in the early morning hours. The vehicle "turned around and drove into the Krantzes' yard, and they could see that there were people up," Lindgren continued. "I imagine there was commotion of all sorts, so this car just left again. It never stopped. They never talked to anybody. He [Willard] assumed it was either that they realized that the fire had been reported or that they had somewhere they had to be at a certain time, or they just plain didn't want to get involved."

As for the children, the *Minneapolis Star* article implies that they had been well taken care of, despite the family's hardship. The newspaper reported that the family had been receiving Red Cross assistance, though the Johnsons were "the least destitute of any that were visited by relief workers."

One piece of potentially incriminating evidence that may or may not have been brought up at the inquest suggested the use of an accelerant at the crime scene, according to local resident Floyd Pinotti. Pinotti knew a thing or two about chasing down bad guys. He spent thirty years in law enforcement, including a stint as Chisago County sheriff. Shortly after he became a deputy sheriff in 1965, the newbie noticed something unusual. "In our evidence locker, which was really a hallway, there was a kerosene can," Pinotti said in a 2018 interview. "I asked what the kerosene can was about. They explained to me that the can was thought to be filled in town with kerosene and brought to the Johnson residence, several hours or several days before the fire." Rumor has it the kerosene can is still collecting dust in an evidence locker somewhere in the county.

Wennerberg didn't mention the kerosene can when he gave an extended statement to the press shortly after the fire. In his remarks, he left no room for doubt that Johnson was still alive and on the run. "We searched every inch of the ruins and we know absolutely Mr. Johnson did not die in the fire," Wennerberg said as quoted in the *Minneapolis Journal*.

Exactly what else might have been learned at the inquest remains a mystery because those records have disappeared. But there must have been enough evidence to convince the authorities that Albin Johnson wasn't just an innocent missing man.

That's because Albin Johnson of Harris, Minnesota, was later charged with first-degree murder in connection with the deaths of his wife and children.

4

"A Hotel, a Livery and Several Stores"

Albin Johnson's hometown of Harris, located in the heart of Chisago County, was first platted in 1873 and was incorporated as a village on July 22, 1884, according to a booklet published in 1984 on the occasion of the town's centennial celebration. The 128-page booklet informs readers that a fellow named Fred Wolf was responsible for a number of firsts in the fledgling village. He was the first settler, the first merchant and the first railroad agent, the publication notes. Perhaps the town should have been called "Wolf, Minnesota." Instead, it was named after Phillip Harris, who is described in the Harris history publication as a "prominent officer of the St. Paul and Duluth railroad." Harris was destined to become a big railroad town.

Those long silver rails came in handy for transporting potatoes, which figured prominently in the village's economy for the first fifty years or so of its existence. Even before the railroad cut through the town, "farmers from near Princeton as well as Cambridge and Grandee brought their potatoes to Harris. It is said that one day, at its marketing height, over 400 loads of potatoes were unloaded," according to the history booklet. In his 1958 book, *Looking Back Over One Hundred Years in Northern Chisago County*, Carl H. Sommer describes Harris as the "Potato Capital of Minnesota." "Before the railroad came through Cambridge, most of the potatoes from that section found their way to Harris," Sommer writes. "At times a whole train was made up with potatoes from Harris. The potato buyers from St. Louis and Chicago made their headquarters in Harris." (Harris also fielded a pretty good baseball team back in the day. But, like Big Hank Johnson after a few too many beers, the games sometimes turned ugly. "I enjoyed many a game against them," Sommer writes, "though many times the game ended in quite an argument.")

By the 1930s, the potato market started to dry up. As the Harris centennial publication notes, "It became more difficult to get good seed, and even with good seed the acreage production was low. Mother Earth had been 'potatoed out.'"

But the town didn't hitch its wagon exclusively to potatoes. Other big cash crops coming and going through Harris included soybeans and corn. Also in its early days, the upstart town boasted "four stores, two hotels, three elevators, three hay presses, two blacksmith shops, a skating rink, an agricultural warehouse, livery stable, two saloons, a meat market, and a depot," the booklet notes. Alfred P. Stolberg and A.O. Stark were among the town's "early merchants and professional men," according to Sommer's book. Both would become prominent figures in the Albin Johnson case.

As of the 2010 census, Harris was home to 1,132 people—good, hospitable, salt-of-the-earth folks, by most appearances. Like so many other sleepy towns across the nation, it has bits and pieces of Americana on display in front of the town's working-class homes. Red, white and blue flags dance in the breeze, while campers and fishing boats stand ready for a trip up north. The town is divided by a set of well-worn, rusty railroad tracks. Folks can truthfully say they grew up on the right or wrong side of the tracks, though it's not immediately clear which is which.

Establishments and attractions along the main drag include a flea market, a bait shop, a post office and a cemetery. There's also a Little League baseball field, a gas station, a storage place and an auto body shop. A Cars R Us business looks out across the street from the obligatory Welcome to Harris sign.

The main street coming into town from Interstate 35 is Stark Road, named after the family of A.O. Stark, the coroner who investigated the Albin Johnson case. The ballfield also bears Stark's name, as do a number of gravestones in the idyllic Covenant Cemetery on the edge of town.

Among the locals is Nan Hult, a silver-haired grandmother with a sharp, curious mind and boundless energy. She lives with her husband, Ron, and the family's cat, Conrad, in nearby North Branch. Swedish and American flags in the home's front yard wave at visitors as they make their way to the front door, which is near a neat and tidy garage where Ron tinkers with his latest woodworking project. A resident of the area since the mid-1960s, Nan

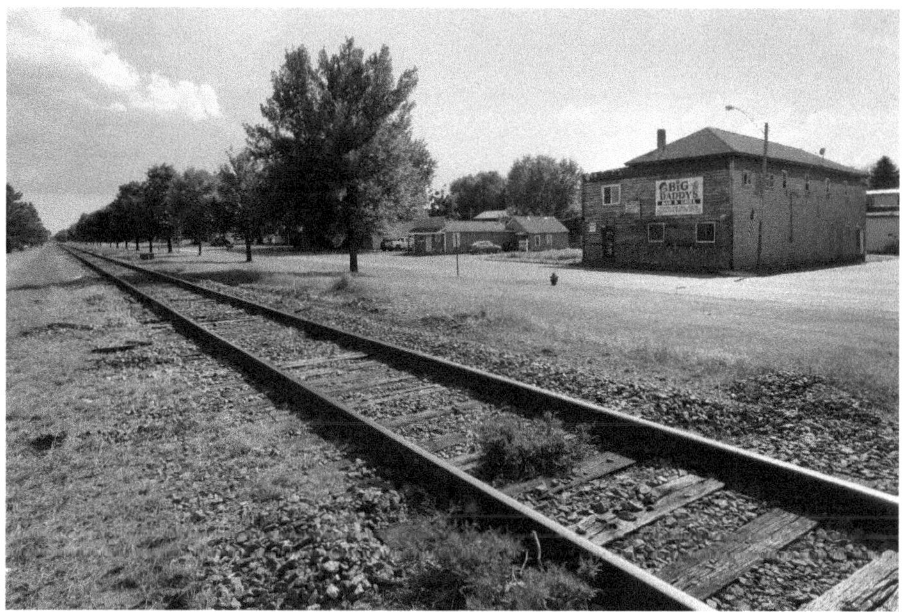

A set of railroad tracks runs through the heart of Harris, with Big Daddy's in the background. *Bill Klotz.*

Hult dove headfirst into researching the Harris fire mystery. Her real name is Nancy Hult, but she may well be called Nancy Drew in a nod to her talent as an amateur sleuth.

During a spring 2017 drive down the main drag of Harris, Hult pointed out the now-shuttered Big Daddy's Bar and Grill. The Big Daddy's building once housed a dance hall on the upper floor and a restaurant down below, she said. Hult heard reports that the bodies of Alvira and her seven children were taken to the dance hall at the future Big Daddy's building, where they were identified by Clayton Anderson, a now-deceased local who was wise to the ways of Harris.

Hult first heard about the Johnson tragedy in the early 2000s, when she and Ron had Anderson over for supper. She "was totally shocked by the horrible tragedy" and started to "delve into it," she says. Anderson "seemed surprised that we hadn't heard about it," Hult added. "He then went on to tell us about it, as he was called by the coroner to come to the dance hall in Harris, where the remains were taken, to be a witness of the handling of the remains."

Another resident from way back, Greg Strom, believes the bodies were brought to the hardware store behind the future Big Daddy's. The store was

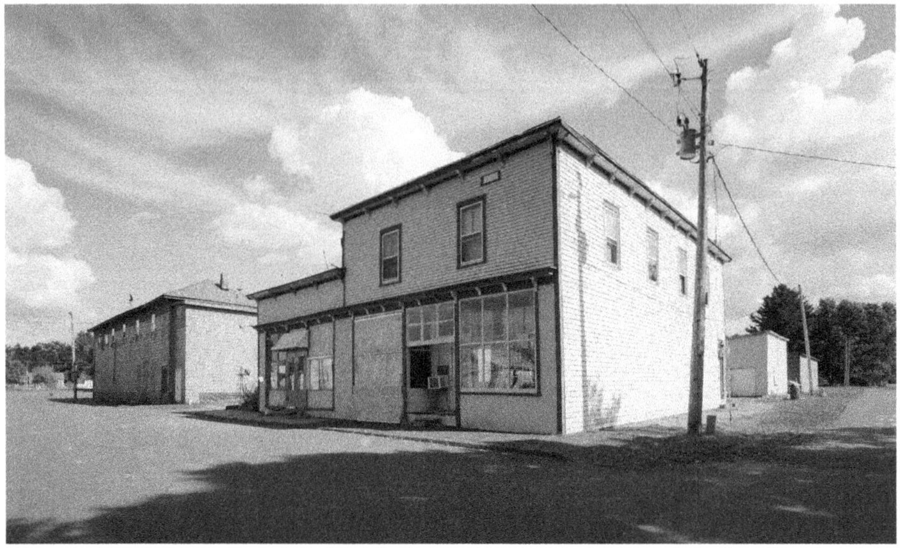

Some locals say the remains of Alvira and her children were brought to this former hardware store, pictured on the right behind Big Daddy's. *Bill Klotz.*

owned in part by A.O. Stark, who was also the deputy county coroner and a mortician.

Back in the 1930s, the hardware store doubled as a makeshift morgue, Strom said. In those days, besides selling nuts and bolts, the hardware store proprietor embalmed bodies and made caskets. These days, Big Daddy's seems like the kind of place that would draw a good crowd when the Minnesota Vikings clash with the Green Bay Packers. The area is somewhat divided in its loyalties to the professional football archrivals, just as it was split down the middle decades ago in the aftermath of the Johnson fire. It's not clear if the name Albin Johnson rang a bell to Big Daddy, but at least one other business in town is well aware of the strange saga of the fugitive farmer.

The Happily Ever After tattoo parlor and the Harris Country Charm Mercantile Shop flank the Kaffe Stuga, a quaint Swedish-themed diner that boasts a steaming hot cup of coffee and a smorgasbord of delicious comfort food. The diner uses Swedish-language signs to let customers know if the place is open (*Vi är öppna*) or closed (*Tyvärr, vi är stängda*). Locals simply call it "the Stuga." A lot of old-timers hang out there for coffee, bright and early in the morning. On a wall of the Stuga, just to the left of the entry, is a wanted poster from the Pinkerton Detective Agency. The poster offers a fifty-dollar

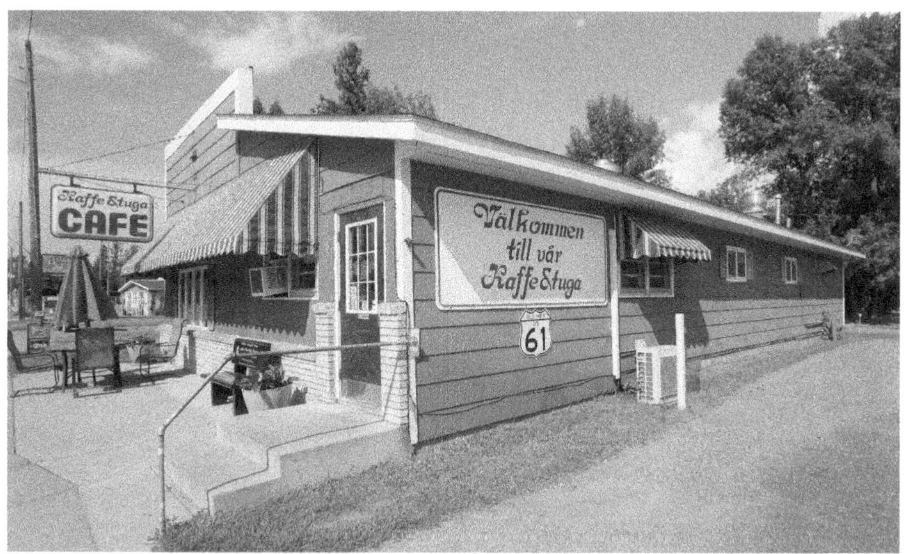

The Kaffe Stuga is a popular hangout in Harris. Albin Johnson's wanted poster is on display. *Bill Klotz.*

reward for information leading to the arrest of Albin Johnson. Fifty bucks was good money in 1933, but nobody was able to claim the booty. It's too late in any case: the offer expired on December 31, 1933.

Still, the publicity generated some fascinating leads. During the first couple of months after the fire, people from multiple states and Canadian provinces claimed to have spotted the missing big man. Albin Johnson findings popped up in Minneapolis, North Dakota, Montana, Pennsylvania, Saskatchewan and Manitoba, among other far-flung locations. One of those false alarms happened in southeast Minneapolis. The *Minneapolis Star* reported on April 27 that Minneapolis police were advised to follow up on a tip that someone bearing resemblance to Johnson had been seen four days earlier in that part of town. Perhaps some of those sightings were hoaxes. Maybe some folks just wanted to get their name in the paper or bask in their fifteen minutes of fame. Others appeared to be good-faith reports from responsible, civic-minded citizens who honestly thought they had seen the missing farmer and wanted to do their part to bring him to justice.

Following up on all leads, the Pinkertons joined the Royal Canadian Mounted Police in searching for Johnson north of the border. Though the agency must have appreciated the tips generated by news coverage of the case, the Pinkertons weren't happy about some information that was fed to the press. That includes bold statements from Albin's brother-in-law Harry A. Galpin, who contended, among other things, that fragments of Albin's bones were indeed found in the ruins of the home.

F.C. Pennington, superintendent of the St. Paul branch of the Pinkerton National Detective Agency, said in a statement to the *Winnipeg Free Press* in early May that Galpin's theories were "handicapping our hunt, since Canadian police naturally would be inclined to let up in their search if they believe such reports are true." Pennington added, "The Canadian police have cooperated splendidly and inasmuch as we intend to continue our hunt in the Canadian provinces we are anxious to halt the circulation of erroneous reports," as quoted by the *Free Press*. Pennington also told the newspaper that three physicians testified at the inquest and that additional examinations of the evidence showed there was no reason to believe that Albin died in the fire.

The Pinkerton agency, incidentally, has come a long way since those wanted poster days. Now based out of an eleven-thousand-square-foot office in Ann Arbor, Michigan, the company is still going strong with one hundred corporate offices worldwide, according to its website. But these days, instead of tracking down train robbers and the Albin Johnsons of the world, the company serves its clients with everything from employment screening to "security risk management." According to its corporate website, the Pinkerton Detective Agency was founded in 1850 by Allan Pinkerton, a Glasgow, Scotland native who made his way to the United States in 1842. Five years later, he joined the Chicago Police Department. After rising to the position of police chief, he hung out a shingle of his own, opening his first private detective office at 80 Washington Street in the Windy City. Throughout the 1850s, the company says, the Pinkerton agency worked with police departments to "arrest criminals anywhere in the country." The agency's reputation soared in 1861, when it uncovered a plot to assassinate President Abraham Lincoln. During the Civil War, Pinkerton headed up the Union Intelligence, from which the U.S. Secret Service was born. In the 1870s, the agency's website boasts, Pinkerton and his agents became "legendary" while in "relentless pursuit" of Jesse James, the Younger Gang, the Dalton Brothers and Butch Cassidy's Wild Bunch. In 1899, the Pinkertons managed to "disband" the Wild Bunch.

But the Pinkertons, for all their guile and experience, were no match for Albin Johnson. "Grandma wanted to find him....She contacted them [the Pinkertons]. But they never found him. That was just money lost, paying them for nothing," said Jeanette Johnson, Alvira's niece.

5
The Lundeens and the Lindstroms

Long before the Pinkertons arrived on the scene, members of the Johnson family—at least the mother and children—did their best to get by in a difficult and unforgiving world.

Alvira Lundeen Johnson's Swedish roots were deep and wide. In fact, her Swedish relatives have been traced to the era just before 1600, when Sweden was a world power. Christopher Hellsing, a doctor who resided in Uppsala, Sweden, was a relative of Alvira's grandmother Anna Stina Hellsing. In a thick hardcover book, Christopher meticulously charted the Hellsing family tree going all the way back to 1590, the birth year of Nils Hellsing. Nils Hellsing had a son named Erik, who was born in 1619 in the province of Värmland, Sweden. Anna Stina Hellsing represented the eighth generation of that branch of the family tree. She was born on February 18, 1833, in Gravendal, Sweden, and was married in 1859 to Johan Jacob Lindstrom.

In the fall of 1885, the Hellsing-Lindstrom clan immigrated to the United States along with their children: Frederick Reinhold, Johanna, Carl Johan (Charley), Jacob Adolf, Gustaf George, Andrew and Anna Kristina, also known as Christine. Another child, Abraham Konrad, died at a young age in Sweden. The family's journey to the U.S. likely began with a horseback ride from their home village, Älvdalen, to Mora. After a short boat ride from Mora to Insjön, they boarded a train to Gothenburg, where another boat was waiting for the long journey across the Atlantic Ocean.

One of the brothers, Carl Johan, worked in Minneapolis for a time. A talented mason, he and his brothers made a living by pouring cement footings around graves. He also helped build Washington Avenue, one of the main streets in downtown Minneapolis. Charley settled into a routine, and he wanted to keep it that way. When his young bride from the home country, Anna, longed to return to Sweden for a visit with family members, Charley refused to let her go. "He informed her if she did go back to her former home, she didn't have to bother to come back," noted Jeanette Johnson, Christine Lundeen's granddaughter. To Anna's credit, she defied her husband's orders. Charley's wife journeyed back to Sweden and never returned. She wasn't among the mourners when Charley Lindstrom died on April 29, 1934, just a year and eighteen days after Alvira Lundeen Johnson and her seven children were found dead.

Another of Christine Lundeen's siblings, Johanna, died in Harris on Christmas Day, a day after Christine's nineteenth birthday. Only thirty-one years old, Johanna died of a hemorrhage while giving birth to a stillborn child, according to cemetery records of the Fish Lake Lutheran Church in Stark, Minnesota. Christine "never talked much about her sister as far as I can remember," said Jeanette Johnson. "The only thing left is a beautiful wool scarf with roses…that my grandmother gave me, saying that it had belonged to her sister Johanna."

Christine took good care of her younger brother, Andrew. They faced a learning curve in adapting to a new country and deciphering a new language, and the young immigrants weren't immune from the discrimination that new Americans still face today. "She had to walk to the church for confirmation. I don't know how far it was," Jeanette Johnson recalled. "And then when they started school, she had to hold his hand and help him. And then they got teased in school. They were called 'greenhorns.'"

The patriarch of the family, Johan Jacob Lindstrom, died on April 9, 1912, in Rush City. The *Rush City Post* hailed him at the time as a "pioneer" in the community. "In Sweden, Mr. Lindstrom was employed in the great iron works of his native country, but followed farming in America, in which he was equally successful," the newspaper reported. "His was a useful life and his character was above reproach. He was a good man, a devoted father and husband, and his life is a noble example to all who knew him."

Anna Stina lived on for eight more years. She suffered a stroke on November 11, 1920, and died four days later at 7:15 p.m., on the seventeenth birthday of her granddaughter Alvira. As the *Rush City Post* described it, the "deceased was conscious until Saturday evening responding to the members of the family at her bedside until Sunday when she answered only feebly and fell into a dreamless sleep, from which she never awakened."

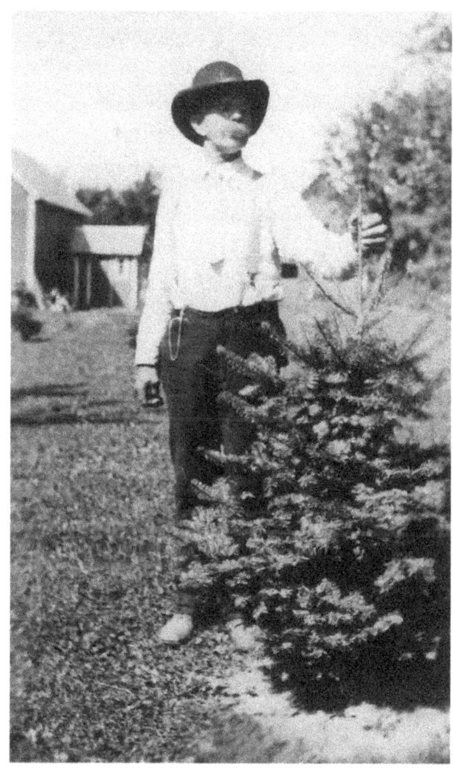

Left: Charley Lindstrom, one of Christine Lundeen's brothers, made a living by pouring cement footings around graves. *Author's collection.*

Below: Anna Stina Hellsing, mother of Christine Lundeen. She's pictured with Christine, *left*, and Christine's sister-in-law, Christina Lindstrom. *Author's collection.*

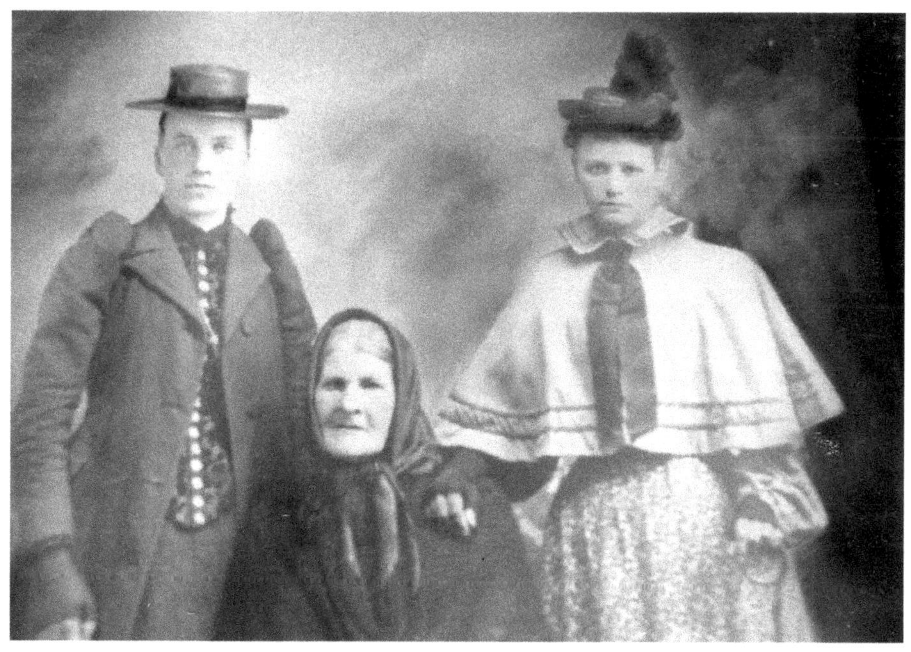

Christine Lundeen broke the news of her mother's passing to her uncle Carl Hellsing, who lived in Söderhamn, Sweden, at the time. Christine was notified when death was near and was "at her side when the end came," she wrote, adding that Anna Stina "breathed her last breath with a song and a prayer."

Before that final prayer, Anna Stina Hellsing loved to sit in her rocking chair in the twilight of her life and dote on her grandkids. The elderly matron of the family would get up to find goodies for Freda, Alvira and the other Lundeen girls when they came for a visit.

Christmas was a special time of year. "One time when Johan and Anna Stina spent Christmas with mother and the family staying overnight, they were awakened early in the morning by a lot of noise. Mother and Alvira were happily discovering what Santa had left during the night," Jeanette Johnson said.

By the time of Anna Stina's death, Fred and Christine Lundeen were well-established members of the community.

Fred Lundeen, like Christine, arrived in the U.S. from his native Sweden at a young age. Mr. Lundeen was only twenty-three when he purchased eighty acres of farmland near Harris from his brother, William Lundeen. A warranty deed, dated July 30, 1885, shows the purchase price was $240, or $3 per acre. Settling onto a site just south of the Emil Johnson property, Fred Lundeen made the most of his investment. He built a sturdy house on the property, farmed the land and raised a family there after marrying Christine in 1893.

"I'm not sure how he and grandma met, but evidently he was considered a good catch, as grandma said she was the envy of all her girl friends," Jeanette Johnson wrote years later. "There isn't much I remember about grandpa, even though years later we lived on the farm with them when my father [Fred Peterson] took over the running of the farm," she added. "I do recall, however, him and my dad reminiscing about the old country and telling ghost stories. They had a good time."

Fred Lundeen was clean shaven and quite handsome in his younger days. When he got older, Alvira's father sported a thick white mustache. "My sister, Betty, remembers grandpa's mustache getting white with the buttermilk he loved to drink," Jeanette recalled.

Fred Lundeen outlived his youngest daughter and her seven children by three years. Before his death in 1936, the aging man had been in poor health for quite some time. Christine Lundeen kept her mind off the loss of her daughter and grandchildren, in part, by throwing herself into a caregiver role for her husband and by leaning on her deep faith.

"I was in high school when grandpa passed away and can remember the casket in the living room of our house," Jeanette wrote. "I had to get permission for time off from school to attend the funeral and had to explain that this meant a lot to me as grandpa was living with us at the time."

Christine Lundeen died on the last day of January in 1972. She was survived by two daughters, eight grandchildren, twenty-five great-grandchildren and eight great-great-grandchildren.

At her funeral, a well-meaning minister did his best to pay homage to a remarkable woman who had endured so much suffering during her 100 years, one month and six days on earth. His job wasn't easy. How do you sum up a century of living in a ten-minute homily? He read some passages from 2 Timothy ("I have fought the good fight, I have finished the race"), he talked about the "very real pain" of "separation" and he assured Christine's mourners that true believers will someday "meet with Mrs. Lundeen again before God's throne." The minister alluded to Great-Grandmother's 100th birthday celebration, which had happened so recently there was probably still some leftover cake to be had. "In spite of the joy in the fact that Mrs. Lundeen was allowed to finish off her 100th year of life, and in spite of the wistful feeling of envy that I personally have felt—and I imagine others have felt—that we might be allowed to see such a tremendous period of history and such a wonderful slice of life and God's creation, there is still an inescapable fact that her life was for all practical purposes finished.... We must face suffering, we must face trial and pain, and we do get older from the day we are born."

He didn't say anything about Alvira and her seven children. He failed to mention that they, too, had "fought the good fight" and finished the race, though their race was more of a sprint than a marathon. And, yes, Alvira and her children "got older from the day they were born." Sadly, they didn't get old.

Christine Lundeen lived out the last years of her life at the Green Acres nursing home in North Branch. A onetime poor farm, Green Acres was dark and scary in the eyes of a kid. The acres didn't seem so green to young visitors. As a little kid, not much more than six years old, I remember cowering in the station wagon with my older siblings when a scary old man pressed his face against the car window. The poor guy was probably harmless, but in a kid's imagination he may as well have been Jack the Ripper.

Ed and Ella Carlson ran Green Acres from about 1935 to 1940, recalls Ralph Carlson, Ed and Ella's son. He, too, remembers it as a disturbing place. "At that time it wasn't just old people," Ralph said after a meeting of the North Chisago Historical Society in June 2017. "It was a lot of mentally disturbed people, too. But they had a player piano. That kept me interested while people were screaming or whatever."

Green Acres is long gone—screaming residents, player piano and all. Now, the former poor farm property is an open, windswept grassy field surrounded by single-family houses for middle-class folks. Green Acres was replaced years ago by one of those modern care facilities in a different part of town. No longer an old folks home, the new place is a senior "community" with fancy "amenities."

About the only reminder of the old place is the poor farm cemetery next to where the building used to sit. As the name implies, the secluded little cemetery functioned as a final resting place for indigent folks who resided on the poor farm. Tiny weather-beaten headstones, engraved with barely readable names, still dot the cemetery landscape in a clearing within a wooded area. The whole place isn't much larger than a tennis court. A few of the stones lie flat on the ground. "The cemetery, we try to protect it from the developers," Carlson said. "I think it's still valid, but there was a lot of unmarked graves there. We have tried to find out who they are. A lot of transient people ended up buried there because there was no other place."

Unlike those visits to Green Acres, trips to the West Coast to visit Christine's daughter, Freda Peterson, were always enjoyable. Freda, my maternal grandmother, was one of four daughters born to Christine Hellsing Lundeen and Fred Lundeen. Her sisters were Olga, Ellen and Alvira.

Olga lived in Minneapolis with her husband, John Zacherson, and their children. The elder Zacherson supported the family by shoveling coal for the railroad—a gig he had for more than fifty years. Olga would come up to the Lindstrom-Lundeen family farm near Rush City in the summertime. "When she came, she always wanted fresh milk, right from a cow," Jeanette Johnson said with a laugh.

One of the Zacherson sons, Bruce, was very close to Jeanette. As an older man, Bruce loved to ride his bicycle on the scenic Cannon Valley trail south of the Twin Cities. He was up there in years, and an unabashed smoker, but he still kept a surprisingly steady pace. Bruce and his riding companion would stop for a soft drink and a buffalo burger at a little beer joint just off the trail. Wearing a ball cap and sucking on a cigarette, he loved to talk about his days in the U.S. Navy during World War II.

Ellen Lundeen Scherer, Olga's younger sister, lived in Rush City with her husband, Matt, and their children. She wore big, round glasses with thick lenses, which magnified the sad eyes that had seen so much hardship.

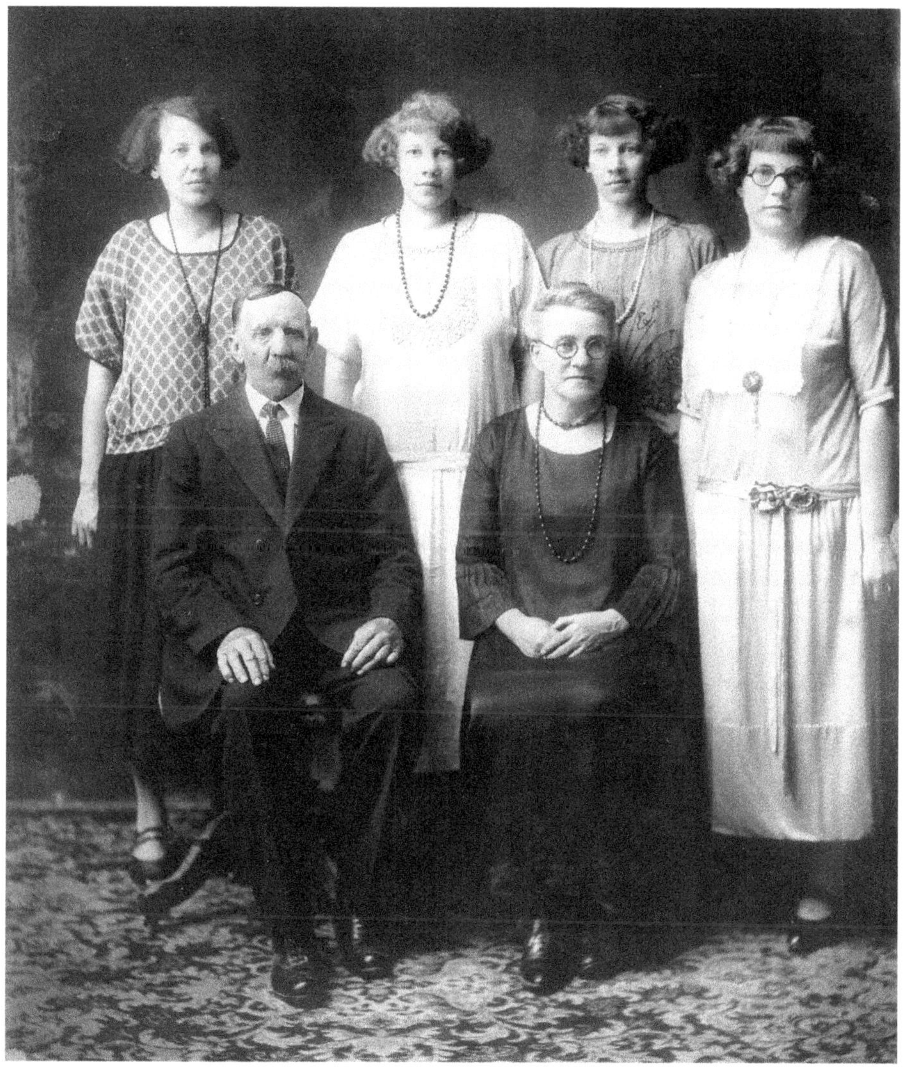

Fred and Christine Lundeen with their daughters—(*from left*) Ellen, Alvira, Olga and Freda. *Author's collection.*

The Harris fire wasn't Ellen's first brush with tragedy and premature death. A few years before the fire, Matt and Ellen's son Richard was struck and killed by a car near the family's home. Richard Scherer was only four years old and is laid to rest at First Lutheran Cemetery, just a few steps from the grave that holds Alvira and her children—and even closer to the final resting place of Henry Johnson, Albin's brother. Affectionately known as

"Dickie," the little boy was hit by the car on November 9, 1929, and he died from his injuries three hours later. Dickie was crossing the pavement when he was struck by a Ford traveling south toward Minneapolis. Three young women in the car "stopped and picked up the boy, and took him to the Grant Hotel, where he was attended by Dr. Holmes," a local newspaper reported. "Later, he was taken to his home, where he passed away. Everything possible was done to save the little fellow's life, but he was injured so badly it was of no avail....He received two scalp wounds, leg broken in two places and other injuries." Funeral services were held at First Lutheran Church in Rush City.

Decades later, at her mother's funeral, Ellen did her best to stay composed. She nervously clutched a tissue as the preacher assured the mourners that "the day is coming when we can meet Mrs. Lundeen again before God's throne."

Matt Scherer, Ellen's husband, died in the early 1960s. In a bizarre coincidence, Matt suffered a fatal heart attack while sitting in the barber's chair, which is precisely how Matt's own father had died years earlier.

Matt was an avid hunter and fisherman, and he knew how to handle a firearm. He passed along his formidable shooting skills to his boys. One son in particular, Ralph, was a crack shot with a rifle. The elder Scherer was "quite a hunter," but "when his son got to be a better shot than he was, he quit hunting," recalled Betty Kollas, Alvira's niece. When he wasn't hunting or fishing, Scherer worked at a flour mill in Rush City. He tried in vain to get his brother-in-law Albin a job at the mill.

After the fire, Matt kept a gun close to his bed. "He thought that maybe Albin had it in for him, because Matt worked at the mill in town and Albin needed a job [and Matt couldn't get him in]," Jeanette Johnson said.

Ellen's younger sister, Freda Peterson, never had it in for anybody—with the possible exception of the annoying guy who peddled appliances on TV and interrupted her favorite shows. Freda and her husband, Fred, lived in Hood River, Oregon, a town that could have inspired a Norman Rockwell painting or a John Denver song. Fruit trees bogged down with cherries, fresh country air and breathtaking views of Mount Hood and Mount Adams fill the senses of all who pass through. A cannery worker for years and a domestic servant in her younger days, Freda was a terrific cook. Her potato salad was world class. She had a quick wit ("Willie Nelson is nice to listen to, as long as you don't have to look at him") and a fun personality. Professional wrestling was must-see television every Saturday night. She never drove, but Fred had a classic car with funky push-buttons instead of a shifting arm to work the automatic transmission.

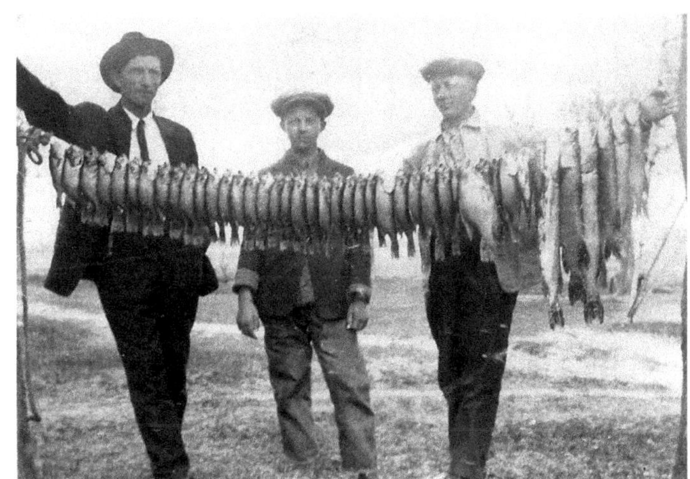

Matt Scherer, *left*, is joined by "Willis" and John Zacherson. Matt reportedly kept a gun by his bed after Albin disappeared. *Author's collection.*

A Swedish immigrant, Fred Peterson ate the same breakfast every morning: oatmeal with coffee as black as midnight. His favorite blue overalls were like a second skin. He wore long underwear year-around, claiming that the undergarment kept him cool in the summer and warm in the winter—like a thermos. Though Swedish was his native tongue, he would sometimes stay awake at night trying to remember a certain phrase in the mother language. His mastery of the language may have faded over the years, but his thick Swedish accent never went away.

Grandma and Grandpa Peterson looked forward to having visitors, including relatives from the old country. Though the visiting Swedes spoke flawless English, Fred and Freda got a chance to brush off their Swedish when the honored guests were in town. Sometimes the conversations morphed into "Swenglish," a combination of Swedish and English.

One thing Grandma Peterson didn't talk about in any language was the untimely death of her sister and seven nieces and nephews. Even into the 1980s and 1990s, she never brought up a subject too painful for her to broach. No doubt, Freda took Alvira's death hard. The sisters were very close in age and in spirit. "They were together all the time," Jeanette Johnson said.

As a child, Alvira had blond hair and a cherubic face. One old photo depicts a young Alvira with long curls resting on her shoulders. She stares into the camera with wide, innocent eyes, oblivious to what the future would hold. "She looked so nice," Jeanette said. "I have a picture of my mother standing straight and tall, thin as a stick. And Alvira is so nice with curly hair. Grandma always had them dress so nice for Sundays and special things: white socks and patent leather shoes."

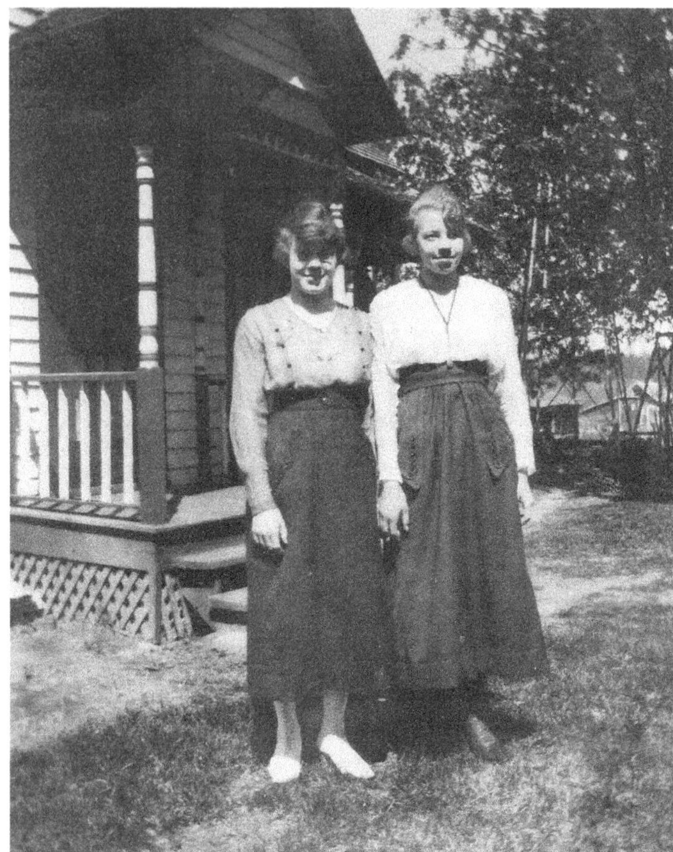

Freda Lundeen Peterson, *left*, and her sister Alvira were very close. *Author's collection.*

Of course, Alvira didn't always dress up so nicely. Like any other farm kid, she enjoyed getting some mud on her shoes once in a while and playing with the animals, including pigs. "My mother said, when they were kids, they had pigs for pets. And they say that pigs make good pets. They are clean, in spite of what you think, most of the time," Jeanette said.

Christine Lundeen was deeply religious, and she saw to it that Alvira and her other daughters were well-schooled in the ways of the Christian faith. Alvira was confirmed in about 1917, while her future husband was working as a lumberjack and farmhand in Canada. In a formal confirmation photo, young Alvira stares at the camera with a pensive look and a hint of melancholy. Sitting in a high-backed chair, she wears a pretty white dress and matching shoes; her hair is tied up in back with a large bow.

Alvira's life wasn't so carefree in later years. Indeed, her adult life must have been difficult with seven young mouths to feed. Naturally, this was

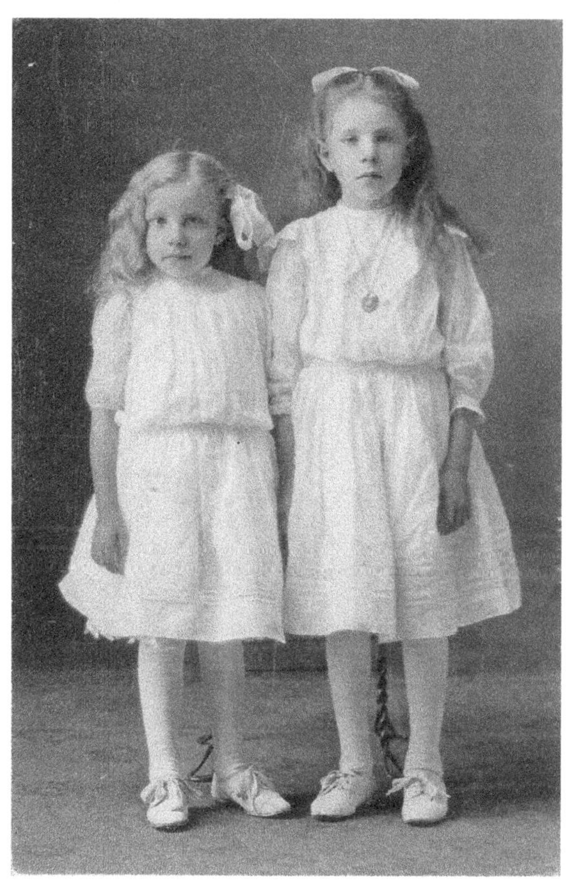

Alvira Lundeen Johnson, *left*, and Freda Lundeen Peterson. The sisters were good playmates as children. *Author's collection.*

long before modern conveniences such as microwave ovens and disposable diapers. Just doing the laundry must have been a full-time job.

As a kid, Jeanette visited the Johnsons from time to time. Alvira's house wasn't especially big, but the family had a barn and plenty of animals, including a pet dog whose remains would later be found in the debris and ashes of the farmhouse.

Much like her sister, Alvira was a good cook. As a kid, Jeanette had a chance to eat at the Johnson place on occasion. And the food was delicious. But Jeanette's visits were preceded by a stern warning from her grandmother that she should not overindulge. "We went over there sometimes to visit because grandma said that we should, you know. But whenever we did go over, she said, 'Don't eat too much, because they don't have much.' But she felt that we should go over there. And the kids, they liked to play too," Jeanette recalls.

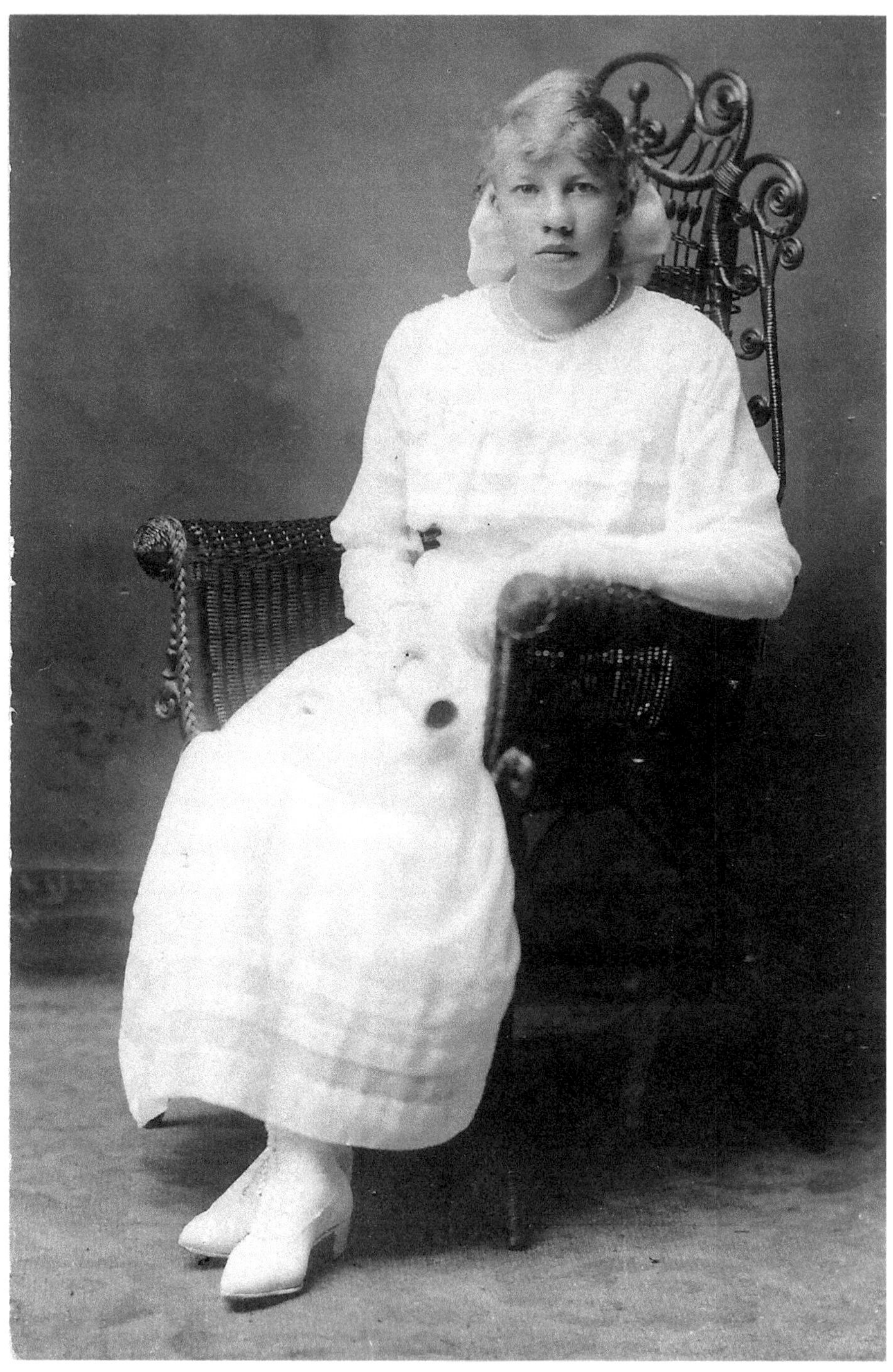

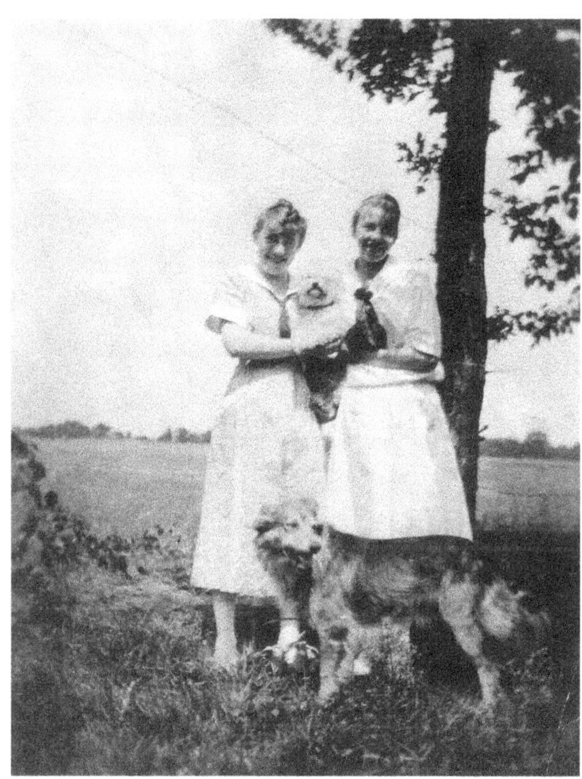

Right: Freda Lundeen Peterson, *left*, and Alvira Lundeen Johnson were big animal lovers. *Author's collection.*

Opposite: Alvira Lundeen Johnson's confirmation photo. *Author's collection.*

What was life like for the kids?

The family was poor, to be sure. It's a safe bet that the children didn't have a lot of toys, given the family's financial situation. At Christmastime, when she was little, Jeanette wondered why Santa Claus never stopped at her cousins' house. But the children did have a few playthings. A faded black-and-white photo shows one of Alvira's boys—perhaps Kenneth or Harold—riding a tricycle. He's wearing a long-sleeved plaid shirt and high-top shoes. The tricycle is one of those nifty, old-fashioned rides, with an oversized wheel out front. He has a firm grip on the handlebars, which are equipped with a handy-dandy bell to let people know that he's on his way. Bike helmets were a thing of the future. Instead, the Johnson youngster is wearing his own special 1930s-style headgear, a tight-fitting lid that looks like a stylistic combination of swimming cap, old-style leather football helmet and winter ski cap. He's cruising past a two-story, wood-paneled house—presumably the same house that would light up the country night just a few short years into the future. He had no way of knowing, of course, that the structure would soon be condemned—not by building code officials, but by bad luck,

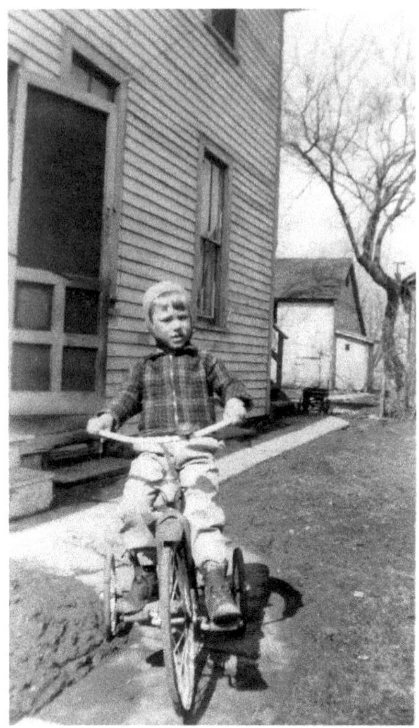

One of Alvira's boys, believed to be Kenneth or Harold, rides past the house on a tricycle. *Author's collection.*

providence, fate or whatever other cruel forces intervened on April 11, 1933. He had no clue that he would meet his death there.

Life on the farm wasn't all fun and games. The kids—at least the older ones—surely had plenty of chores to do. Down on the farm in those days, every able body was expected to pitch in. But they also had pets, including at least one dog, and lots of open space in which to run around and play games. There was plenty of fresh country air to breathe and places to explore for children with abundant energy.

In one undated photo, three of the brothers stand in front of a large bush with their cousin Bruce, from the big city. A dog sits patiently with the boys, looking pleased in front of the camera. On the far left is Clifford, who stares back at the photographer with sad, puppy dog eyes. A beaming Harold stands beside him. To Harold's left is their cousin Bruce. Kenneth, on the far right, is dressed in overalls. With his head cocked slightly to the side, he gives the photographer an inquisitive look, as if to say, "I'm bored with all of this picture-taking. Let's grab a fishing pole and get us some bluegills."

In another photo, Clifford and Dorothy are pictured with their backs against the house. They must have had the sun in their eyes, because both kids are squinting. Clifford is dressed like a typical farm boy in overalls that don't fit quite right. Dorothy, wearing a flowing white dress, smiles broadly at the camera, seemingly happy in spite of the family's hardship. A closer look at the photo offers some clues about the family's financial situation. Put another way, it's safe to say they didn't get to the store very often. Both kids' shoes are scuffed and tattered. Clifford's are particularly battle-worn, with gaping holes in front.

The school-age Johnson kids attended classes at Chippewa Hill School, a classic one-room structure that offered basic education for kids in grades one through eight.

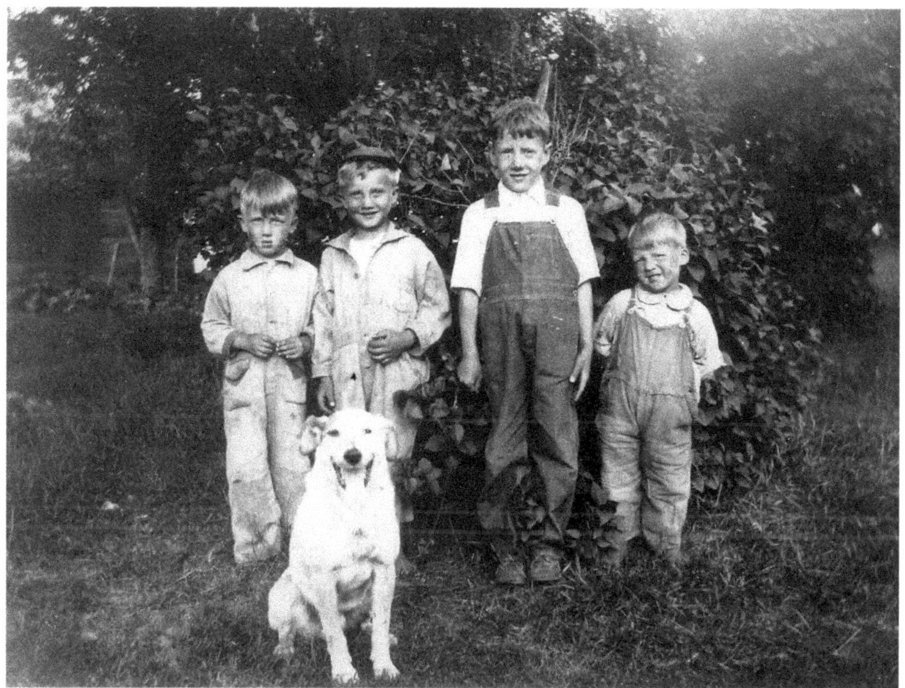

Three of Alvira's boys and their cousin from the city, Bruce Zacherson. From left are Clifford, Harold, Bruce and Kenneth. *Author's collection.*

Betty Kollas offers some insight as to what school was like for the Johnson kids. She attended classes at Chippewa Hill in the 1930s and 1940s, shortly after the Johnsons were there. The school was about a mile away from Betty's childhood home, which was near the Johnson place. If the kids were lucky, they hitched a ride on a horse-drawn wagon. But most of them walked, regardless of the weather conditions.

"We walked the driveway and then we walked on another road with a slope that connected to what I knew as the Government Road, because it was kept up by the government," Kollas said. In the winter, "I would walk to school and my eyelashes would get icicles on them," she remembers. "Mom would want Dad to take me to school, on horses, I suppose. And I said, 'Nope, I'm going to walk. Nobody else has a ride to school.' I wasn't going to be a sissy.'"

The schoolhouse certainly wouldn't compare with today's $100 million schools, which are equipped with theaters, flexible learning spaces, security guards and big athletic fields with press boxes. But by the standards of the day, Chippewa Hill School was actually rather big and quite nice. "Basically,

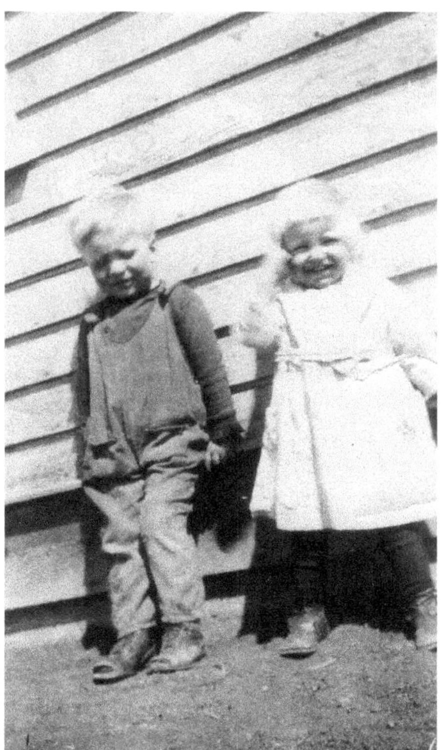

Right: Clifford Johnson, *left*, and Dorothy Johnson in front of the house. *Author's collection.*

Below: The Peterson family. From left are Freda, Fred, Betty and Jeanette. Fred Peterson initially believed that Albin's body would be found in the ruins of the home. *Author's collection.*

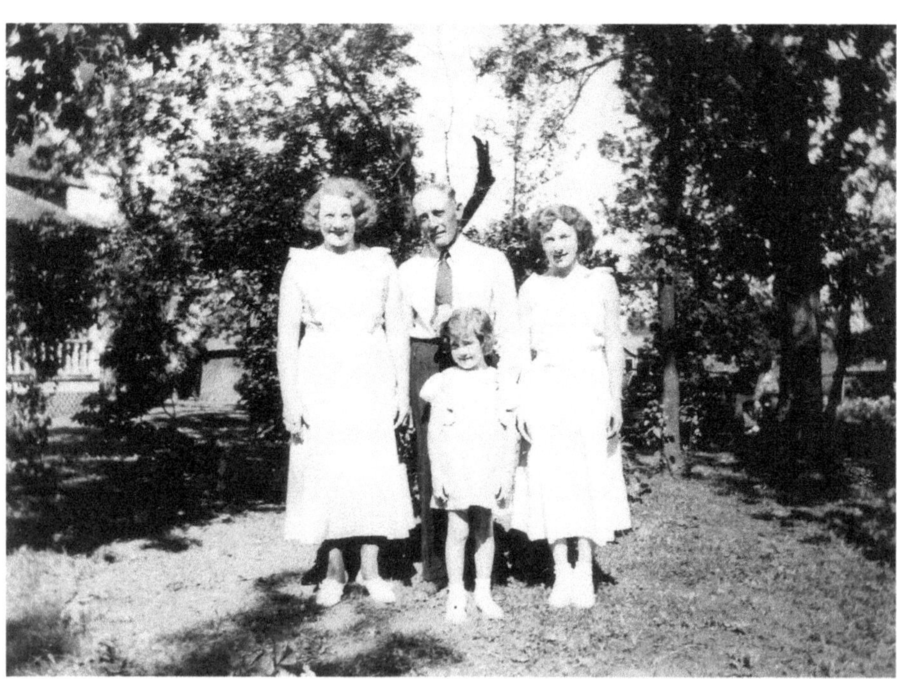

it wasn't that bad. You could almost call it a 'progressive school.' If you had gone through the first grades and kept learning, you saw and heard everything. So by the time you were in eighth grade, you had a well-rounded education," Betty said.

The school had trouble finding teachers to work in the country, Betty recalls. The underpaid teachers typically lived with local farmers near the school. The teachers did their best to stock the building with supplies and equipment.

Inside the single room were a library, an entry, a small kitchen and a cloakroom, which was filled at times with the powerful aroma of onion sandwiches brought to school by some of the less-fortunate kids. Maybe the Johnson kids were among those who had to settle for meatless lunches, though perhaps they occasionally enjoyed chicken or beef from the animals raised on the farm.

During recess, Chippewa Hill students played baseball or softball on a field in front of the school. "I got orders that, 'When the ball comes your way, you hang onto it,'" Betty said. "And I could catch that ball pretty good. And my fingers got so swelled up I couldn't even make a fist. It's a wonder I didn't get arthritis."

Firewood, a broom and other cleaning supplies were stocked in the "fuel room" to keep the place warm and tidy. The schoolhouse also had a "great big stove" and a "big chimney across the whole room," Betty recalls. A young scholar who lived close to school was asked to arrive early during the wintertime to stoke up the fire and make sure the place was nice and toasty. "Sometimes it would get so hot it would turn red. It's a wonder it didn't burn down," Betty recalled. "But it kept the school pretty warm."

Of the seven Johnson kids, at least three were old enough to attend school: ten-year-old Harold, nine-year-old Clifford and seven-year-old Kenneth. Five-year-old Dorothy was probably too young to attend classes in those days. Then there was four-year-old Bernice, two-year-old Lester and four-month-old James.

Mrs. Lundeen was especially close to Harold. Perhaps she had a soft spot for the eldest son because she saw three of her own boys die in infancy in the late 1800s and early 1900s.

Christine's boys—Elof, Adolf and Edwin—were buried in an unmarked grave at Taylor Cemetery in Rush City. Johan Jacob Lindstrom, the boys' grandfather, was placed in the ground there before his remains were moved to First Lutheran Cemetery in Rush City.

Christine never talked much about the boys or the loss of Alvira and her children. Perhaps the grave was unmarked because the family couldn't afford a nice piece of granite engraved with the boys' names and words of comfort from the Bible. Or maybe Christine didn't want to be reminded of her loss. Years later, Betty and Jeanette purchased a gravestone inscribed with the children's names and birth dates so future generations could pay their respects.

As for the Johnson kids, Jeanette Johnson imagines that life must have been difficult for them and their mother. Not just because they were poor, but because Alvira and the kids had to put up with Albin. "What a life she must have lived with that man," she said. "He couldn't have treated his family very well."

After the fire, the entire community drifted between shock, bewilderment and mourning. Hundreds of mourners packed into the small Lutheran church in Rush City for the funeral of Alvira and the children. Chippewa Hill School was shut down for the memorial service.

A story that ran in the Rush City newspaper covers the funeral in a surprisingly matter-of-fact way. Under an all-caps headline that reads

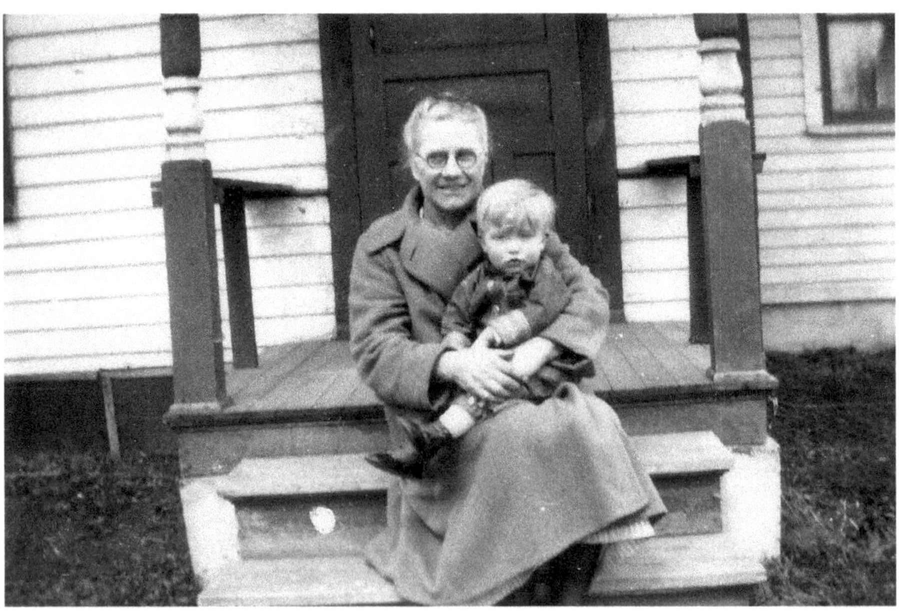

Christine Lundeen and her grandson Harold. *Author's collection.*

"LAST RITES FOR MOTHER-CHILDREN," the story reports that "about 350 people crowded the First Lutheran Church to witness the last solemn rites pronounced in the burial of the victims of one of the most terrible tragedies that has ever occurred in this community." The obituary continues:

> *The eight in one flower-decked casket were borne to the First Lutheran Cemetery, where they were interred in the family lot belonging to Mr. and Mrs. Fred Lundeen, parents of the mother.* [Alvira] *was confirmed in the First Lutheran Church at Rush City and married to Albin Johnson, son of Mr. and Mrs. Emil Johnson of Harris. Seven children were born to this union: Harold, age 10; Clifford, age 9; Kenneth, age 7; Dorothy, age 5; Bernice, age 4; Lester, age 2; and James, age 4 months.*

After informing readers that the "music was rendered by Mrs. Arthur Nelson and Mrs. Einar Larson, who were accompanied by Mrs. Chas. C. Wilson," the story gets around to saying a few nice words about Alvira. "Mrs. Johnson was a gentle character whose love and care for her children and home spoke of the untiring and courageous disposition, which were her chief characteristics."

The story dances around the topic of the fugitive husband/father, even though a manhunt for Albin Johnson was in full swing by April 15, the date of the funeral. Still, the spirit of the missing farmer hung over the mourners like yet another dark cloud that was about to burst.

Jeanette Johnson recalls that Albin's name came up at least once during the memorial service. "I remember the service. The church was packed—packed," Mrs. Johnson said. "The preacher announced that they were looking for Albin Johnson."

The *Rush City Post* article barely mentions the seven children, but Jeanette Johnson hasn't forgotten her young cousins. She still gets emotional when talking about the children who were taken away far too soon. She has a black-and-white photo of a group of students from Chippewa Hill School, herself included. With a shock of blond hair, a young Harold Johnson sits in the front row.

Jeanette, close in age to Harold, recalls playing with him all those years ago. The kids exchanged valentines and often visited each other. In a sense, Jeanette filled the role of the big sister that Harold never had. "We had a wagon that somebody made for us, and I used to pull him in the wagon. He

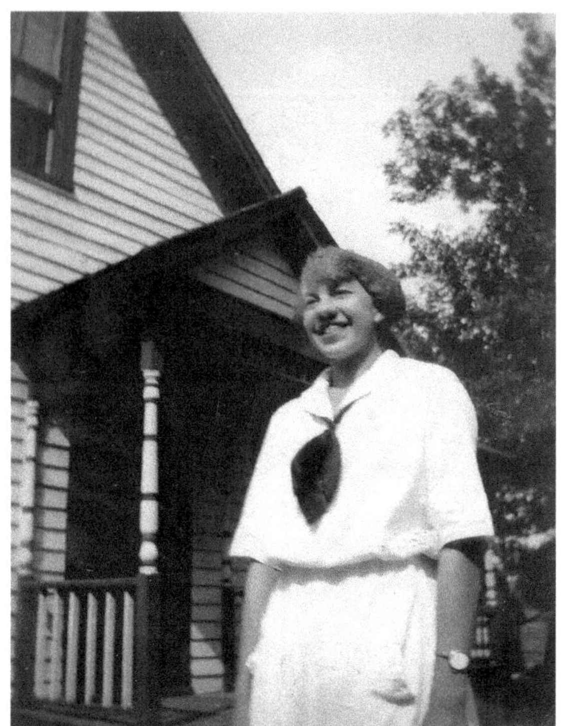

Right: Alvira Lundeen Johnson in front of the house. *Author's collection.*

Below: Chippewa Hill School. Harold Johnson is sitting in front, far left. Cousin Jeanette is fourth from the left in the back row. *Author's collection.*

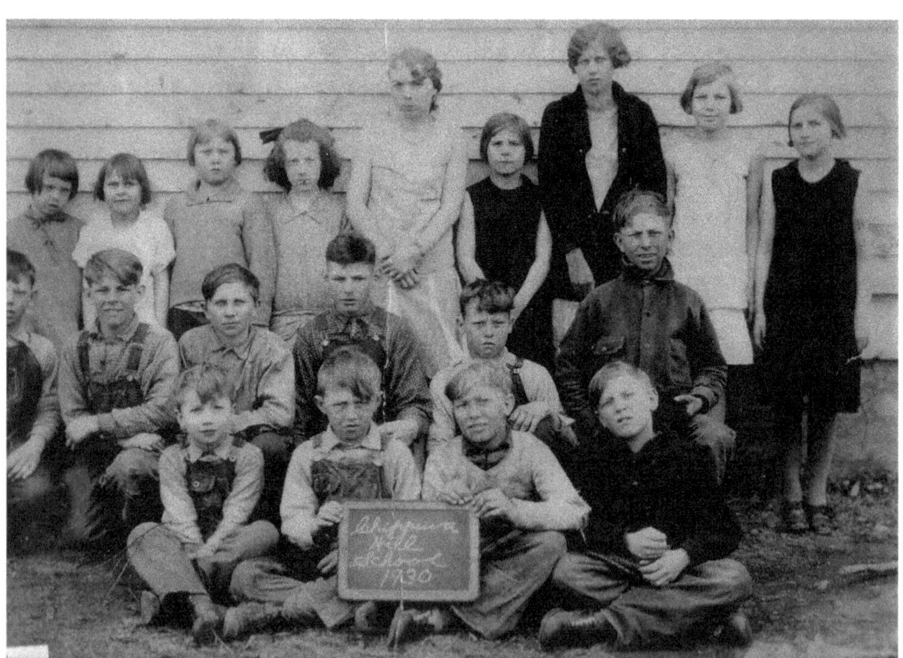

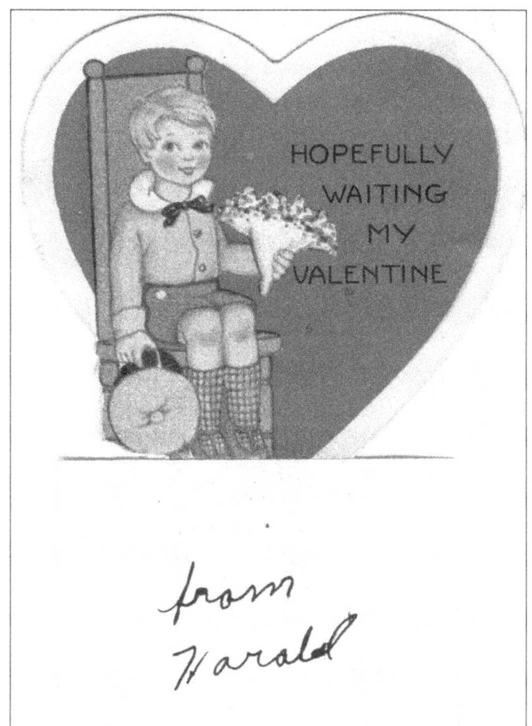

Left: A valentine from Harold Johnson to his cousin Jeanette. *Author's collection.*

Below: Jeanette Peterson Johnson pulls her cousin Harold Johnson in a wagon. *Author's collection.*

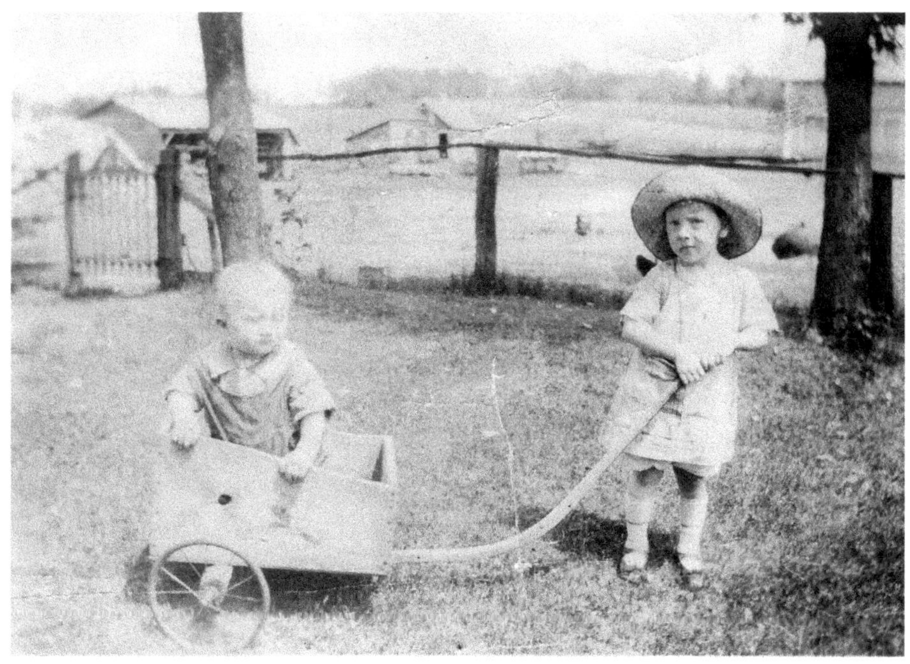

was younger than me," she said. Then her mind drifts back to the tragedy. "Oh, that was terrible," she said. "What a shock that was to the whole area there....The church was packed with people for that funeral. All of them in one casket. All of them."

6

A Town in Mourning and a Posse of Farmers

On April 18, just a week after the fire, a crowd of people gathered in and around the area that used to be home to Albin and Alvira Johnson and their seven children. Nearly half the town of Harris showed up, it seems, along with residents of Rush City and other nearby villages. Some of the gatherers, no doubt, were mere adventure seekers who wanted to be in on the action. Perhaps a morbid sense of curiosity lured them out of their comfortable homes to the soggy, hilly terrain near the ruins of the Johnson place. Others—including big shots from St. Paul and Chisago County—truly believed they would hit pay dirt with a clean sweep of the countryside. Even the chilly water of the mighty St. Croix River was on their radar as they set out on their mission. The searchers all had one goal in mind: to find something—anything—that would lead them to the fugitive farmer.

Leonard J. Lund, the state's deputy fire marshal, led the search, according to the April 18, 1933 edition of the *Minneapolis Journal*. As the newspaper put it, "A posse of more than 300 farmers from Harris and neighboring communities was formed and a thorough search of the woods and swampland in the district began." The searchers covered a wide swath of land from Harris to the St. Croix River, while "river-men" combed the St. Croix River from Sunrise to the "ferry near Rush City, a distance of six miles," according to the *Chisago County Press*.

By late April, conditions for searching the choppy waters of the St. Croix were ideal. The river was getting shallower by the day. On April 27, it was about two feet lower than it had been at the time of the fire, the *Chisago*

County Press noted. But the river didn't yield any secrets and the searchers came up empty.

At about this time, the *Minneapolis Journal* presented a bizarre theory: the idea that Albin may have been injured, but not killed, during the fire. Those who clung to that belief must have had a hard time convincing skeptics how an injured Johnson could have eluded the hundreds of searchers, who were traipsing through the countryside trying desperately to track him down.

The Johnsons were too poor to own a car or a truck, and any sod-busting they did was presumably accomplished with the help of horses or mules. There's no indication that Johnson commandeered a horse from the old homestead or hitched a ride, unless you believe the mysterious tire tracks left at the scene were made by a getaway vehicle.

Aside from that unlikely scenario, it's safe to assume he had to make tracks on foot. Perhaps an injured Albin Johnson could have hobbled a mile or two before collapsing. And yet, no trace of Johnson—dead, alive or debilitated—was to be found.

On April 27, the *Chisago County Press* offered readers a detailed description of Albin and how he was likely dressed. He is described "as having worn a blue Scotch cap, dark suit, pair of overalls over his trousers rolled halfway up to his knees, and black shoes. He was 43 years of age [contrary to other reports that he was "about 46"], 6 feet and 3 or 4 inches tall, weighed 230 pounds, had dark hair mixed with gray and was smooth shaven. His complexion was described as being dark." No one could escape the glowering image of the man with the blue Scotch cap. Jeanette Johnson remembers seeing wanted posters plastered up all over town and in surrounding communities. His dour mug was everywhere and nowhere, all at the same time. "Albin's face was posted in all the cafes and...buildings there. When we stopped in North Branch, we saw a picture. Sunrise, that was another little town [with a poster]. Wanted," she said.

Even with Albin's cold eyes staring back at patrons all over town, nobody came forward with any helpful clues. And reporters increasingly turned their attention to other stories.

The news reports eventually started to slow down, though an update on the still fruitless search appeared in the May 18, 1933 edition of the *St. Paul Dispatch*. The story leads with the news that authorities were "more bewildered than they ever were" but that they "trusted...time to yield a clue to the tragedy."

For the first time in any published reports, a relative of Albin Johnson's spoke up and defended the missing man. Johnson's seventy-one-year-old

ROBERT A. PINKERTON, PRESIDENT DAVID C. THORNHILL, ASS'T TO PRESIDENT ASHER ROSSETTER, VICE PRES & GEN'L MGR
RALPH DUDLEY, VICE PRES. & ASS'T GEN'L MGR. O. C. TURRELL, SEC'Y & TREASURER
H. S. MOSHER, MGR. DEP'T CRIMINAL INVESTIGATION

PINKERTON'S NATIONAL DETECTIVE AGENCY, INC.
(FOUNDED BY ALLAN PINKERTON 1850)

OFFICES

ATLANTA	CLEVELAND	HARTFORD	MILWAUKEE	PITTSBURGH	SCRANTON
BALTIMORE	DALLAS	HOUSTON	MONTREAL	PORTLAND ORE.	SEATTLE
BOSTON	DENVER	INDIANAPOLIS	NEW ORLEANS	PROVIDENCE	SYRACUSE
BUFFALO	DETROIT	KANSAS CITY	NEW YORK	RICHMOND	ST. LOUIS
CHICAGO	HARRISBURG	LOS ANGELES	PHILADELPHIA	SAN FRANCISCO	ST. PAUL
CINCINNATI					TORONTO

$50.00 REWARD

The above amount is offered by a client of this Agency for information leading to the whereabouts of

ALBIN JOHNSON

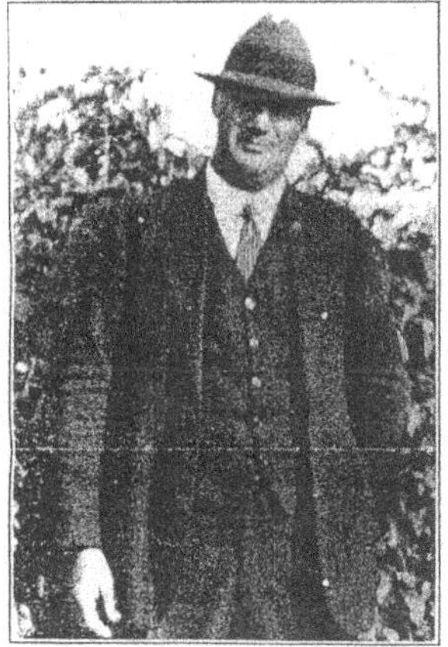

DESCRIPTION

Age 43 years; height 6 feet 3 inches; weight 240 pounds; hair dark, turning gray; blue eyes. Large hands, very strong, typical farmer and woodsman. Blue overalls over dark suit and woodsman's cap.

He disappeared from Harris, Minnesota, April 10, after his home was destroyed by fire and his wife and seven children were burned to death. He has worked in Saskatchewan, Canada, and may again attempt to enter Canada. Has little money and may seek food and shelter at relief stations.

Officers on duty in northern woods are particularly requested to be on the look-out for this man.

Should any information as to this man's whereabouts be obtained, kindly notify the nearest Pinkerton office by wire or telephone at our expense; or Sheriff James Smith, Center City, Minn.

Pinkerton's National Detective Agency, Inc., under its rules does not operate for rewards, therefore will not accept, nor permit any of its employes to accept this reward or any part thereof.

Pinkerton's National Detective Agency, Inc.

360 Robert Street Telephone Cedar 0855 or (after hours) Nestor 3277 St. Paul, Minn.

April 20, 1933

THIS REWARD OFFER EXPIRES DECEMBER 31, 1933, AT MIDNIGHT.

The Pinkerton National Detective Agency offered a fifty-dollar reward for "information leading to the whereabouts of Albin Johnson." *©Pinkerton Consulting & Investigations, Inc. 2016. Used with permission.*

father, Carl A.E. "Emil" Johnson, and other family members cited the unsuccessful search effort as proof that the younger Johnson must have been dead. Emil Johnson "pointed to lack of a trace of Johnson after the fire to back their belief that he too perished in the flames," the *Chisago County Press* reported. Johnson didn't explain how it was possible to find bone fragments of every other family member, including the four-month-old baby, James, while at the same time finding no trace of Albin's six-foot, three-inch, 230-pound frame (or 240 pounds, as some reports had it).

Still, some folks who knew Albin weren't ready to convict him just yet. Fred Peterson, one of Albin's brothers-in-law, seemed to agree with Emil Johnson, at least initially. He was quoted in a *St. Paul Dispatch* article from April 12, just a day after the fire. "I am sure that continued search of the ruins in the basement, the only part left of the entire house, will yield the body of Johnson," said Peterson, whose wife, Freda, was one of Alvira's three sisters.

Another brother-in-law, Harry Galpin, was more adamant. Insisting that Johnson died in the fire, Galpin claimed that the authorities botched the investigation and perhaps even hindered neighbors from putting out the blaze, as noted in the April 28, 1933 edition of the *Minneapolis Tribune*. Galpin claimed that farmers who arrived at the scene with water in milk cans were "denied permission to put out the fire" and that the authorities failed to establish a fire line "to keep out the hundreds of persons who swarmed all over the place and thus could destroy any remaining evidence of the missing man," according to the *Tribune*. Galpin also dismissed reports that a man resembling Albin had been seen in Minneapolis since the fire. There are "at least 1,000 men in Minneapolis who would answer his general description," he said, as reported by the *Minneapolis Tribune*.

Albin Johnson sightings in the U.S. weren't confined to Minneapolis. In early May, news reports surfaced that a man bearing Johnson's description was seen in Reading, Pennsylvania—a curious development considering that Albin had no known connections to that part of the country. Daniel Acison, a Pennsylvania man who had seen pictures of the missing man in the newspaper, told Reading police that a man who looked a lot like Johnson was spotted in the Reading railroad yards on Sunday, April 30. It turned out to be a false alarm.

Meanwhile, the sense of urgency to find Johnson was growing. By late April, the authorities were working under the assumption that Johnson was a family annihilator. In May, news surfaced of a potential murder weapon or weapons. Wennerberg said that "two pistols and a rifle found in the ruins may have been used to slay the victims."

But that physical evidence didn't bring the searchers any closer to finding Albin. Wennerberg sounded increasingly frustrated. By May, he "admitted temporary failure toward solving the mystery" and reiterated that Johnson might have escaped the fire and "drowned himself in the nearby St. Croix River, or fled to Canada, where he once lived," the *Dispatch* reported. Still, "there's not much to go on," Wennerberg admitted. "But we'll press ahead and maybe time will yield some tangible clue."

Wennerberg, grasping to find some reason for hope, recalled another baffling crime that had tested his skills as a sleuth thirteen years earlier. In that case, a local farmer was found dead in his barn, an apparent murder victim. The killer didn't leave any clues. The case seemed hopeless. But Wennerberg pressed on, and his persistence paid off handsomely. Six weeks after the lifeless body of the unfortunate farmer was discovered, Wennerberg tracked the killer down, arrested him and "sent him to prison for life," the *Dispatch* reported.

As the 1933 case evolved from a fire investigation/missing person inquiry to a full-blown murder investigation, Wennerberg probably longed for the day he would see big Albin in a prison cell. By April 26, Alvira Johnson and her seven children were officially murder victims. At that time, the University of Minnesota's Dr. Charles A. Erdmann said in a report that the victims "apparently had been poisoned or beaten to death where they slept," according to the *Brainerd Daily Dispatch*. That was all Wennerberg needed to hear to get off the fence. Based on that information, Wennerberg said flat-out that the mother and children were murdered—and that the suspected killer was on the loose. The investigator asked the Royal Canadian Mounted Police and immigration officers at the Canadian border to join Minnesota authorities in keeping a lookout for Johnson.

On the U.S. side of the border, Dr. Erdmann didn't shed much light on the question of Albin's whereabouts. But he had a lot to say about the fate of the other family members. If Dr. Erdmann got it wrong or somehow misinterpreted the evidence, it wasn't because he lacked experience. By the spring of 1933, Dr. Erdmann was wrapping up his fourth decade of service in the field of medicine.

A bald man with round, wire-framed spectacles and a prominent nose, Charles A. Erdmann was born in Milwaukee, Wisconsin, on August 3, 1866, just a year and four months after General Robert E. Lee surrendered to General Ulysses S. Grant. In 1896, three years after he earned his medical degree from the University of Minnesota, Erdmann married Caroline Edgar. The couple had two sons and one daughter.

When he wasn't practicing medicine or spending time with the family, Erdmann was active in the Masonic lodge movement. He was a thirty-third-degree mason, a "past master" of the University Lodge No. 316 in Minneapolis and a member of the Zurah Temple, according to his obituary in the *Minneapolis Star*.

Erdmann, who also studied at the University of Vienna, "joined the medical faculty of the University of Minnesota in 1894 where he served 42 years without missing a day until he retired in 1936," the *Minneapolis Tribune* reported on February 19, 1941, at the time of his death.

Three years before his retirement, Dr. Erdmann made a strong case that Alvira Lundeen Johnson and her seven children were murder victims. Dismissing Galpin's theory that Albin was among the victims, Dr. Erdmann was confident that only eight bodies were found at the scene of the fire. Dr. Erdmann pinpointed where the bodies were in the ruins—and why they were presumed to be dead when the house went up in flames. "Five of the youngsters had gone to sleep on mattresses on the floor two rooms away from the kitchen where the blaze presumably originated," Erdmann said as reported by the *Winnipeg Free Press*. "Another slept in another part of the house and the mother and an infant slept in the dining room next to the kitchen.... It would be natural to assume that, granting the fire started in the stove in the kitchen, that at least some of the five children sleeping two rooms away in the living room would have been awakened by smoke, assuming they were alive, and would have had an opportunity to move about. Instead, the bodies of all eight persons were found in the same positions in which they went to sleep." Based on a "study of the bodies and circumstances surrounding their discovery," Erdmann concluded that the mother and children died "by poison or some other manner" prior to the fire, according to the Associated Press as reported in the *Albert Lea Evening Tribune*.

In early May, not long after Erdmann's conclusions were reported in the press, there appeared to be a major break in the case. Authorities followed up on reports that a man resembling Johnson had been seen north of the border, where he was believed to have been hiding out. It seemed to make sense. Johnson had waxed nostalgic about the old days in Canada, long before he had a wife and seven young children to fret about, not to mention an empty bank account. Wennerberg told reporters that Johnson had "expressed a wish to be back in Canada" just a day before the fire.

After the fire, Johnson was believed to have crossed the border into Canada at Walhalla, a port of entry on U.S. Highway 32 at the border of North Dakota and Manitoba. The Royal Canadian Mounted Police believed they were hot on Johnson's trail. The *Brainerd Daily Dispatch* reported that authorities were "less than 24 hours behind" a "road-stained traveler believed to be Johnson....He was last seen at a farm house near Morden, Manitoba, six miles across the international boundary, Friday morning (May 5)....Making his way on foot in territory strange to him, it was not believed he could have progressed far."

At one point, Johnson supposedly stopped at a farmhouse to get some food. "The suspect told the farm family from which he begged breakfast that he had crossed the border during the night," but the stranger was hesitant to give further information and "said little else," according to the *Dispatch*. The family members notified authorities, explaining that the hungry caller matched the description of the fugitive from Minnesota. It all seemed to make sense. The pieces, at last, seemed to be falling neatly into place.

"Furthermore, it was pointed out, if the suspect is Johnson, he is doing exactly what Minnesota authorities had anticipated—making his way back into Canada, where he worked as a ferry hand thirteen years ago and of which he had spoken with fondness as recently as three days before the tragedy," the newspaper reported.

The *Winnipeg Evening Tribune* picked up on the story in a big way. On the morning of May 5, readers of the newspaper poured a cup of coffee, rubbed the sleep out of their eyes and read an all-caps headline across the front page of the paper: "ALBIN JOHNSON TRACED TO MANITOBA." In the subhead, readers learned that the accused killer was seen near Winkler, Manitoba, and that Royal Canadian Mounted Police rushed to the town in the southern part of the province to follow up on the lead. According to the story: "It was reported to police at [Morden, Manitoba] that a visitor from Walhalla, Minnesota, had seen a man answering the description of Johnson in the vicinity of Haskett, Manitoba, just north of the boundary near Walhalla."

In addition, "word came from St. Paul that S.B. Wennerberg, county attorney, Center City, Minnesota, had received word from a detective that a man answering Johnson's description ate breakfast at a farm south of Winkler, Manitoba, Thursday morning." The farmer took a gander at a photograph of Johnson and promptly declared he was the man. "The county attorney did not say who made the report to him, but it is known that Detective Mason, of the Pinkerton National Detective Agency, was working in the vicinity of Haskett, Manitoba," the newspaper reported. Wennerberg was advised that the mystery man, after having breakfast, "begged a ride from another farmer and accompanied him toward Winkler."

Wennerberg told the newspaper he was confident they were "on the right trail." "Johnson formerly lived in Saskatchewan, and I believe he'd head toward Canada," Wennerberg said. The farmer who claimed to have served breakfast to Johnson explained that the mystery man was wearing a "blue suit, gray cap and brown shoes, which tally with the attire of Johnson the day before the fire," the newspaper reported.

The reported sighting was broadcast on Canadian radio at the request of the Royal Canadian Mounted Police. Circulars and wanted posters were distributed throughout the area. Johnson was described as a "typical farmer and woodsman, with large, strong hands." Inspired by the big tip and hoping to make an arrest, the officers of the law brought all hands on deck. "All available police constables of the southern territory reported at Morden to assist in the search, and all residents of the area have been asked to keep watch of the suspect," the Winnipeg newspaper continued. But that hot lead quickly turned cold, as did another reported Johnson sighting.

On May 6, the *St. Cloud Times* reported that a man believed to be Johnson was picked up near Cavalier, North Dakota, after having crossed the international line from Manitoba. The man "aroused suspicions of authorities when he crossed the international boundary twice in two days. Sheriff A.B. O'Connor of Pembina County said he would be held for United States immigration authorities for crossing from Canada into this country without permission," the newspaper reported. The man taken in by police turned out to be thirty-four-year-old Michael Joseph Coen, who claimed he was from Ireland and had relatives in California, according to the *Winnipeg Free Press*. Coen told authorities he had "wandered from place to place and crossed the international boundary on several occasions," the newspaper reported. Coen was jailed for a time while officials did a background check. His story checked out.

Meanwhile, investigators in Canada advanced the theory that Johnson had taken his own life. The *Winnipeg Tribune* reported on May 6 that authorities expected to find Johnson's dead body in their search of the "whole countryside," including the "coulees and brush" between the towns of Winkler and Morden, Manitoba.

The searchers suspected Johnson had been traveling in the southern district of the Canadian province. They were following up on a tip that he had hitched a ride from one of the locals, presumably in the direction of Prelate, Saskatchewan, which is about 590 miles straight west of Winkler. The authorities assumed Prelate was his final destination because Johnson had once worked there as a farmhand. But they didn't expect to find him alive.

"With the whole countryside on the alert for the wanted man, police believe that if it was Johnson who rode to Winkler with a farmer Thursday morning he has realized the futility of attempting to escape and committed suicide," the *Tribune* story continues.

Meanwhile, the *Journal* added, searchers were following up on clues in both countries. On the afternoon of May 5, police in Portage la Prairie were informed that Johnson had been seen "prowling through the bush to the south of the city." That compelling lead turned out to be a case of mistaken identity. The Royal Canadian Mounted Police made an "intensive search of the vicinity," only to find that the man believed to be the farmer from Harris was "a Portage man resembling Johnson."

Less than a month after the fire, a sense of resignation was already falling over the community, which was making the transition from shock and disbelief to hopelessness.

Under a headline that read "FRIENDS VIEW JOHNSON CASE WITHOUT HOPE," the *St. Cloud Times* on May 10, 1933, captured the somber mood of the community while at the same time acknowledging that facts were hard to come by. Folks in town were reluctant to say anything when members of the press showed up with pens and notebooks in hand, the *Times* reported. "Neighbors of the Johnsons, scattered over the desolate, hilly countryside, tightened their lips [and] shook their heads when a reporter appeared," the newspaper notes. "Their guarded comments revealed that the Johnsons were just like any other rural folks. There was a big family, a hard struggle to pay rent on the old Johnson homestead, owned by Carl A.E. 'Emil' Johnson, father of Albin. But father and mother apparently were devoted to their offspring."

"There had been some fresh grief in the family for he had been unable to meet rental payments on his father's farm and had planned to move. Fred Peterson, who like Johnson, married one of the Fred Lundeen girls, told about this. Then came the fire."

Emil Johnson was, understandably, too shaken up to shed much light on the mystery, other than to repeat his contention that Albin had died in the fire, the newspaper reported.

Leaving out the fact that Emil had evicted Albin and family, the *Times* reported that Emil Johnson was "living alone in a two-room cottage" and was "scarcely able to discuss the tragedy" or the circumstances behind Albin's "decision" to leave.

Painting a vivid word picture, the article notes that Emil's body "trembled" and his words were "distorted with grief" when he tersely told the newspaper that he didn't know "what could have happened."

Still, as local legend has it, at least one Harris resident who had accompanied the authorities to Canada in search of Johnson had a very good idea of what happened.

Enter Paul Peers into the story.

7

What Did Mr. Peers Know?

Paul Knowles Peers came into the world on July 18, 1887. The son of Mr. and Mrs. Archibald Peers, Paul grew up on a farm in Harris and went on to serve in France during World War I. He later worked for the Chisago County Highway Department and the Amber Milling Company in Rush City before retiring in 1952. When he was only eighteen, Peers made his way north. Perhaps he was looking for a bit of excitement or an opportunity to work in the logging camps. In any case, he spent about three years in Saskatchewan, Canada, before returning home to Rush City. Given his experience in Canada, it's no surprise that the authorities tapped Peers to be part of the Albin Johnson search party north of the border. Exactly what he saw or heard up there went with Peers to his grave when he died in 1963. But rumor has it he encountered Big Albin up north, contrary to conventional wisdom that Johnson was never seen or heard from again after the fire.

Years ago, Hult spoke to Peers's daughter about the incident. As the daughter remembered it, the elder Peers swore "up and down that when they went to Canada they saw Albin," Hult said. "And I said, 'That's not possible, because they went up there to get him [and would have arrested him]. Why would they not get him?'"

Greg Strom, who grew up near Fish Lake, did a report on the Johnson tragedy for a school project in the early 1970s. As part of his research, young Mr. Strom interviewed Peers's daughter. "She was pretty sure her dad saw Albin. But he didn't bring him back because he didn't have jurisdiction in Canada," said Strom, whose grandfather, Albert Strom, ran the general store in Harris for many years. Or perhaps he was afraid to bring Johnson to justice out of concern for his family. Everyone knew that Albin and his brothers were big, rough men, and some say they were prone to violent outbursts. "Maybe when Albin was in Canada, he threatened the family," Strom speculated. "Albin was physically intimidating and his brothers were physically intimidating. When his brothers were still around, people were afraid to talk about it—and afraid of pursuing Albin." Strom theorizes it's highly possible that folks who may have had information years ago were intimidated into keeping quiet. Even eight decades later, many old-timers are loath to talk about it. Strom didn't hear a word about the Johnson tragedy for decades after doing that initial school report. And he was rebuffed in his efforts to mine old-timers for information. After approaching one elderly resident who remembered the fire, Strom was firmly told to "drop it."

Even in 1933, people were more than happy to let it go and get on with their lives. With nobody talking and few solid leads, the story went cold again and the newspaper reports dried up for a few months. It wasn't until October, nearly six months after the fire, that the next big development in the Johnson saga took place. On October 3, 1933, the case went before a grand jury, presided over by Judge Alfred P. Stolberg, a well-known Chisago County resident.

Stolberg was an imposing, stern-looking character with round spectacles and a receding hairline. He had the look of a man who relished the chance to mete out justice. No stranger to the bench, Stolberg was a second-generation judge in East Central Minnesota. Born on December 4, 1876, he was the son of the Honorable Peter Stolberg, a Chisago County judge, and the former Sarah Larson. The family settled in Harris, according to his obituary. Following in his father's footsteps, the younger Stolberg graduated from Gustavus Adolphus College in St. Peter and went on to earn his law degree at the University of Minnesota. After a stint in the office of the register of deeds, he served in the Chisago County Attorney's Office and then accepted a judgeship in the Nineteenth Judicial Court District, which is where his path crossed with the ghost of Albin Johnson.

Official records of the grand jury proceedings, assuming they still exist, have never been released, despite efforts from researchers to pry them loose. (In the fall of 2018, a records clerk for the Chisago County Court Administration wrote to the author that an archives search of the grand jury proceedings came up empty.) But enough evidence was presented to the court to satisfy Stolberg that Johnson hadn't died in the fire—and that he very likely had blood on his hands.

In the fall of 1933, a grand jury indicted Albin Johnson in absentia for first-degree murder. On October 3, "when the grand jury met" in Chisago County, "the Johnson case was laid before the jurors," the *St. Paul Dispatch* reported that month. "The [secret] indictment, it was reported, was returned at that time," though Johnson's relatives "steadfastly maintained that he, too, died in the fire," the newspaper reported.

With that, the husky farmer from Harris, Minnesota, was officially a murder suspect—at least in the eyes of the grand jury.

But Albin Johnson was never brought to justice. The unsolved mystery begs a host of questions. If Albin Johnson had been apprehended, would he have been convicted of killing his wife and seven children? Was there enough evidence to convince a jury of his peers that he committed the crime? The evidence suggests that a crime had taken place, given the position of the bodies at the time they were found and the fact that Johnson was the only one who escaped the vengeance of the flames, according to the authorities. Certainly the odds would have been against Johnson, who didn't have the means to hire a seasoned attorney. And it would have been easy for the prosecutor to find sympathetic ears on the jury after presenting gruesome evidence of a young mother and seven children dying under suspicious circumstances.

Moreover, his rights were limited. In 1933, the U.S. Supreme Court hadn't yet ruled that an accused individual has a right to counsel, said David Shultz, a law professor at Hamline University in St. Paul and a visiting professor at the University of Minnesota Law School. That wasn't settled by the high court until later in the 1930s, he said. "It's possible, if he couldn't afford it, he might not have had an attorney right off the bat....And not until later, in the 1960s [with the Miranda ruling] does the court say you have to be advised of that right....So it's very possible that he could have been arrested, interrogated, not told about his right to counsel. Maybe he confesses, maybe he's put on trial and he may not have an attorney to represent him," Schultz said. "It's possible that in a high-profile case like that the court might have found someone to volunteer [to represent him], but those are a couple of scenarios."

What about evidence? To the extent that forensic evidence was available in 1933, what could the prosecutors have presented to the jury? Naturally, DNA evidence didn't exist back then and the science of fingerprinting was a work in progress, so it's questionable what could have been brought into the courtroom, Schultz said. "Certainly the stuff we see on *CSI*—none of that would have existed at that point," he said.

On the other hand, a heap of other evidence would have been fair game. Defendants had "limited protections in terms of the exclusionary rule," to the point where police "could bring in almost any evidence," including that which was illegally obtained, Schultz said. Chances are, judge and jury would have rendered a guilty verdict, and the fugitive farmer would have faced life behind bars. (If he had been captured and convicted, Johnson would have escaped the gallows. Minnesota abolished capital punishment in 1911 after a botched hanging.)

Albin Johnson would not have been the first person desperate enough to kill his family. Such crimes happen more often than may be expected. And sometimes it comes from the most unlikely sources.

Standing just over six feet tall and weighing 180 pounds, William Bradford Bishop Jr. spoke five languages—English, French, Italian, Serbo-Croatian and Spanish. As of 1976, the Pasadena, California native worked as a U.S. Foreign Service officer. He liked to drink scotch and wine, and he had a taste for spicy foods. A true outdoorsman, Bishop liked to fish, swim, golf, play tennis, ski and ride motorcycles. He loved to read. He was educated at Yale and received a master's degree in Italian from Middlebury College in Vermont. But he also battled depression, was self-absorbed and prone to violent outbursts. And the feds are convinced he became a family annihilator.

The above biographical information is from Bishop's wanted poster. He was added to the FBI's "Ten Most Wanted" list on April 10, 2014—precisely eighty-one years after the Harris fire. He was removed from the list in June 2018 without ever having been captured, according to the *Wall Street Journal*. His story is eerily similar to that of Albin Johnson. On March 1, 1976, in Bethesda, Maryland, Bishop allegedly killed his mother, his thirty-seven-year-old wife and three sons, ages five, ten and fourteen, with a blunt instrument. He then allegedly transported the remains to Columbia, North Carolina, "where he buried the bodies in a shallow grave and lit them on fire," according to his wanted poster. In a 2014 press release, the FBI said that Bishop was last seen one day after the murders at a sporting goods store in Jacksonville, North Carolina. He bought a pair of sneakers there.

"Nothing has changed since March 2, 1976, when Bishop was last seen except the passage of time," Steve Vogt, special agent in charge of the FBI's Baltimore Division, said in the 2014 press release. According to the release, investigators extensively searched the area where the bodies were discovered and found no sign of Bishop. "When Bishop took off in 1976, there was no social media, no twenty-four-hour news cycle," Vogt said. "There was no sustained way to get his face out there like there is today. And the only way to catch this guy is through the public." Vogt added, "If Bishop is living with a new identity, he's got to be somebody's next-door neighbor....Don't forget that five people were murdered. Bishop needs to be held accountable for that."

Thirteen years before the Harris fire, reporters jumped on another odd case involving a mysterious fire and a missing man. In that case, E.J. "Ed" Sailstad of Eau Claire, Wisconsin, was believed to have died on August 27, 1920, in a fire at his cottage. Call it the "Is Ed Dead?" mystery.

Unlike the Albin Johnson case, the narrative surrounding Sailstad, a father of two young children, reads more like a soap opera script than a real-life murder mystery. Citing a "web of circumstantial evidence," the *Minneapolis Star* reported in October 1920 that Sailstad allegedly staged his death and then ran off with his stenographer, Dorothy Anderson. Evidence to that effect included testimony from a Duluth taxi driver who claimed he had driven Sailstad and his mistress from Lake Nebagamon to Duluth after the events of August 27, the newspaper reported. Adding a salacious detail to the story, a hotel worker in Superior claimed that the alleged lovebirds spent the night together at the hotel on the night of the presumed deadly blaze.

Sailstad's wife, identified only as "Mrs. E.J. Sailstad," wasn't buying it. Foreshadowing a claim that would be made years later by Albin Johnson's defenders, she was convinced that her husband had indeed died in the fire. "I won't believe that Ed is alive and did not burn to death in that cottage until they bring him to me, or prove he is a prisoner," Mrs. Sailstad said, as quoted in the October 16, 1920 edition of the *Minneapolis Star.*

Unlike Ed Sailstad, Albin Johnson didn't have a stenographer—or a mistress, at least as far as anyone knows. But like Bishop, he may well have been depressed and prone to violent outbursts. He certainly had reason to be in a bad state of mind. Given his dire situation—evicted from his home by his father with little or no hope for gainful employment during the Great Depression—he may very well have been driven to madness.

Despite the circumstantial evidence against him, Johnson had his supporters. The most vocal was Harry A. Galpin, his faithful brother-in-law.

8

Albin Johnson's Biggest Defender

Harry A. Galpin was born in Rochester, Minnesota, on January 4, 1884—exactly six years and one day before his future brother-in-law came into the world. He married Olga Johnson, one of Albin's three sisters, on September 15, 1912. Galpin's death certificate lists his occupation as a salesman. It's not clear if he had any legal training, but in the years after the fire, Galpin took it upon himself to be Albin's informal legal advocate—at least in the court of public opinion. Albin would have been hard-pressed to find a more loyal brother-in-law.

Three years after the fire, Galpin fired off a scathing affidavit in which he defended Johnson's honor and blasted the authorities' handling of the Harris fire investigation. From the judge on down, no one escaped his wrath. He accused Deputy Coroner A.O. Stark, Sheriff James A. Smith and County Attorney S. Bernard Wennerberg of making "not less than 31 errors" ranging from destruction of evidence to negligence. Specifically, Galpin alleged that Stark had "cast [Albin] Johnson's remains" outside the foundation of the fire-ravaged home, where the remains were "trampled under foot by spectators." He intimated that the judge, the Honorable Alfred P. Stolberg, was part of an old-boys club that cared little for the truth and was only interested in self-promotion. As Galpin saw it, the whole lot was incompetent and the search for Johnson was a waste of taxpayer money.

Galpin's affidavit rambled on for a dozen pages, and much of it is conspiratorial in tone and content. Still, he made some good points. Noting that Albin had only a sixth-grade education, for example, Galpin reasoned that the missing man wasn't smart enough to pull off a crime and disappearing act that even a seasoned criminal would have been hard-pressed to commit. Galpin opens his October 1936 affidavit by declaring, quite accurately, that the "fatal Albin Johnson fire" is "still an unsolved mystery." He claimed that the grand jury indictment of Johnson was based "on pure speculation." Galpin wrote, "Three and a half years have passed since the night of the fire, yet not a single trace of Johnson has been found, not a particle of evidence he has ever been alive since the fire. Even the strange automobile tracks leading in and out of the farm gate have been satisfactorily explained away." Galpin didn't elaborate on the last point, so it's hard for the reader to judge if, indeed, there was a satisfactory explanation for the tire tracks.

He went on to defend Johnson as a man with an "enviable reputation among his neighbors" and a "giant of a man," a characterization that many would dispute. Johnson had no money or experience in chicanery, Galpin notes. And yet, the authorities claimed that he "planned his crime so carefully that the state can find no motive that will stand up under cold, logical analysis." They failed to prove that the fire was set intentionally or that Johnson was seen alive after the fire, he said. Galpin suggested that Johnson's bones were in fact found in the ashes and debris of the farmhouse, despite assurances to the contrary from the powers that be.

On May 4, 1933, under a headline that read "NEW EVIDENCE FOUND IN FIRE," the Associated Press reported that Galpin "brought back about 50 fragments of bones which he said he discovered at the scene of the fire." The fragments were under the ruins of the pantry, about fifteen feet from the remains of other family members, he said. Galpin claimed he also found a piece of melted glass. Why was that significant? "Glass melts at 2500 degrees Fahrenheit," Galpin told the Associated Press. "It simply demonstrated the terrific heat at the location at which these bones were found and may explain why larger remains were not disclosed." Expanding on that story, Galpin said in the affidavit that a deputy sheriff and unidentified relatives uncovered the human bones "in a distant part of the fire ruins, six days too late to do Johnson any good." Galpin turned the bones over to Dr. Erdmann, the University of Minnesota professor of anatomy and the state's scientific expert in the case. He also notified Wennerberg, the county attorney. But the authorities were unconvinced.

Covering all his bases, Johnson's biggest defender summarized a number of hypothetical explanations. For instance, Big Albin might have "stayed away from home that night" or "escaped with his clothing ablaze," or perhaps he was somehow "prevented from being there that night," Galpin wrote. Other scenarios: Albin "lost his mind, killed his family and fled," or perhaps he was "so badly cremated his remains were unrecognized by the inexperienced volunteer searchers." Galpin added, "Careful analysis shows that time and [searching] have eliminated all but the last possibility, for this sixth grader [*sic*] was not a brilliant man. He would have been picked up within 48 hours."

Not content to stop there, Galpin also raised doubts about Erdmann's scientific opinion that Albin had not died in the fire. In Galpin's view, Erdmann had his blinders on when it came to investigating the case. He charged that Erdmann's theories were based more on speculation than science. "At one time Dr. Erdmann...stated 'he believed' the unexplained bones found in a distant part of the ruins [Albin's bones, in Galpin's view] were those of the children," Galpin wrote, continuing,

> *Mind you, he did not state that as a scientific fact but merely as his belief, founded on other presumptions he thought were facts.... We found there were only four possible ways that children's bones could have gotten where these bones were found. Each of these possibilities were explored and one by one rejected as absurd or impossible under the conditions known to have existed here–therefore they could not have been children's bones. They must have been the remains of the father.... We asked the doctor to give us a signed statement that he could not identify these bones, but he replied he "did not know what use we intended to make of such a statement, and besides, we had not hired him."*

Next, Galpin turned to the alleged errors in the investigation. "Cold, logical analysis shows that Albert O. Stark, deputy coroner, Harris, Minnesota, James A. Smith, sheriff, Lindstrom, and S. Bernard Wennerberg, county attorney, Center City, committed not less than thirty-one (31) errors in their handling of this one case." Now Galpin really takes off the gloves, as he accuses nearly everyone involved in the case of being crooked, inept or untruthful:

> *These charges range from perjury and destruction of evidence to incompetence and negligence, and to railroading a murder indictment against a man known to be dead, to covering up the ghastly blunder of the deputy coroner who had cast Johnson's remains outside the foundation where they were trampled underfoot by spectators.... Cold, logical analysis also shows the state's scientific witness (Erdmann) turned his back upon science and*

> *based his conclusions upon perjured testimony, and upon six (6) rebuttable presumptions of fact.*

Galpin then took a shot at Stolberg, the district judge in Center City. In Galpin's view, the judge "proved himself unfitted to fill his high and honorable position." Judge Stolberg lived about four miles from the Johnson farm and was a stockholder and board member of the State Bank of Harris, Galpin noted. He was tight with Stark, also known in Galpin-speak as the "bungling deputy coroner."

That's where it gets interesting, at least in Galpin's view. Stark was the president of the bank, and the judge's brother, G.J. Stolberg, was a cashier at the bank (and the survivor of at least one robbery attempt, as previously noted). The late E.W. Stark, the coroner's brother, served for a time as state treasurer.

Why were the judge's ties to the bank important? "The judge was intimately acquainted with many of the prominent characters in this mystery, and as the so-called 'murder case' was the biggest 'crime' that had ever happened in Chisago County, it was to be expected the judge was deeply concerned with the lack of progress in the case," Galpin wrote. "He had to know a great deal about it, since it might be necessary for him to approve expenditures of county funds in the so-called 'search' for Johnson." Connecting the dots, Galpin saw it as a twisted conspiracy and a desire on the judge's part to protect the deputy coroner "against loss to his reputation and to his many business connections."

Galpin couldn't resist the urge to taunt the authorities and ridicule them over their quixotic search for the missing man. Galpin similarly lambasted the Harris fire investigators. He said sarcastically that the private detective and deputy sheriff returned from their "four thousand mile joy ride" and "fake search" with "long noses and longer faces, but without a particle of new evidence." In a more serious vein, Galpin wondered why the judge called a regular grand jury instead of a "special grand jury under a special prosecutor." He noted the judge's alleged unwillingness to hire an orthopedic surgeon to examine the bones to determine "if they were from a child or an adult." Galpin wrote, "The judge is well aware of his own errors and must find it difficult to keep from laughing at himself while dealing out justice from the bench, but if we can prevail upon him to take his tongue out of his cheek long enough to answer these very puzzling questions, he will make us happy beyond words." Referring to what he called the "deception and conspiracy" of authorities and investigators, Galpin opined that the coroner, sheriff and county attorney "developed an astonishing inability to think" and had their minds made up in advance. Those overseeing the case had "from the first come to believe a crime had been committed, and the amazing fact that there never has been any evidence of a crime mattered little to them," Galpin charged.

At one point, Galpin tried in vain to get Minnesota governor Floyd B. Olson to clear Johnson's name. He also tried to get the legislature involved. His end game was to have the murder indictment annulled and a burial certificate issued for Albin Johnson. Governor Olson didn't want to have anything to do with the case, which comes as no surprise to David Schultz, the Hamline University and University of Minnesota law professor. One consideration: the fire had occurred only three years earlier, and it was quite possible that Johnson was still alive. Though the case had been cold for some time, it's a good bet that Wennerberg and company still held out hope of tracking down the missing farmer and putting him on trial. "Rarely do governors want to step into proceedings like this, especially when they see no political benefit," Schultz said in an interview. "A governor stepping into a murder trial like this might actually cause problems in terms of pretrial publicity, prejudicing the trial. So I think most of the time they stay out of it."

After coming up empty, Galpin embarked on a new strategy: publicly ridicule the judge and Albin's accusers until he was sued for libel, "when we could then bring out evidence into court." That would eventually come back to bite him. Galpin's affidavit described the inquest as a "farce" and the grand jury investigation as a "joke." In an over-the-top closing statement, Galpin wrote:

> *The bloodiest of Kentucky's mountain feuds were started with less provocation than this. A thorough investigation of this whole case under a judge unbiased by his personal opinions with witnesses examined under a lie-detector would seem to be the best solution....But that would be impossible because the only way you can ever get any of the county officials or the doctor there would be to bring them hand-cuffed and leg-ironed.*
>
> *The complete story of the way the officials mishandled their duties... in this one case is the blackest chapter in the record of Minnesota jurisprudence. And if this railroaded murder indictment is not annulled, every proud, law-abiding citizen of Minnesota is made a party to it whether they like it or not.*

Galpin put a lot of passion into his defense of Johnson, but he failed to explain why the investigators would railroad Albin Johnson, except for the unproven assertion that they were trying to protect the business interests of Stark and company. Simply put, they had no reason to conspire against a lowly farmer.

Unlike Johnson, Galpin himself was eventually brought to justice—not for murder, but for allegedly besmirching the good names of his adversaries. On October 14, 1939, Galpin was convicted of libel against Judge Stolberg and other county officials. He was sentenced to two six-month terms in jail. Pleading guilty, Galpin admitted to putting up posters in public places that accused Stolberg and other officials of being "ignorant, incompetent, negligent or crooked" and of hounding Johnson to cover up their own "ghastly blunders," according to the *Minneapolis Tribune*.

Judge Joseph J. Moriarty of Shakopee ordered Albin's beleaguered brother-in-law to spend three months at the Washington County Jail in Stillwater, with the remaining time to be suspended if he "has a record of good behavior," the *Tribune* reported on October 15, 1939. The jail sentence must have been a bitter pill for Galpin to swallow, and it may well have hastened his death. On February 17, 1940, at the Bedford, Indiana home of a "Mrs. Garrett," Harry Asa Galpin suffered a coronary thrombosis and died an hour later, according to his death certificate and obituary. He was far away from the bucolic countryside of Harris, Minnesota, and the elusive ghost of Albin Johnson.

S. Bernard Wennerberg, a favorite target of Galpin's vitriol, outlived Galpin by nearly thirty-seven years. The onetime pursuer of Albin Johnson died in relative obscurity on January 9, 1977; this was twenty-five years to the day after the passing of his Swedish immigrant mother, Anna Anderson Wennerberg. He was eighty-seven. A paid obituary in the *Minneapolis Star* suffices to say that Wennerberg "passed away at the Parmly residence in Chisago City." He was buried at Fairview Cemetery in Stillwater, Minnesota.

As a postscript to the news of Galpin's sentencing, the *Minneapolis Tribune* reminded readers that Johnson was still a fugitive. "The six-year mystery of Johnson's whereabouts remains unbroken," the newspaper reported.

9
Gangsters and Hoodlums

When it comes to crime stories, the Albin Johnson case has a lot to offer. But the case had plenty of competition for the public's attention back in the day. The 1920s and early 1930s were rife with lurid tales of gangsters and hoodlums running roughshod over the laws of decent society. Outlaws of the era included the likes of Ma Barker, of the rowdy Barker-Karpis Gang, and gangster John Dillinger, who was reputed to have run amok on an extensive crime spree that included helping himself to cash on hand at twenty-four banks.

But no scoundrels or scalawags captured the headlines or the public's fascination more than the infamous boyfriend-girlfriend duo of Clyde Barrow and Bonnie Parker. Bonnie and Clyde achieved almost cult-like status—not just in the U.S., but throughout the world. As recently as 2017, Swedish Radio ran a documentary on the outlaws. According to the radio show, "They were lovebirds who dreamed of freedom and a better life. And they would do just about anything to live the kind of life they wanted to live."

On April 13, 1933, just two days after a mysterious blaze turned the Johnson farmhouse in Harris into a ghastly crematorium, Bonnie and Clyde appeared to be trapped. "Outside, the house was surrounded by police—lots of police," Swedish Radio reported. "It was a sufficient force for regular criminals, but not enough for people who didn't have anything to lose.... When the police ordered them to come out with their hands over their heads, they got their answer in the form of gunfire. A gunfight broke out. Clyde was hit in the chest, but it was just a surface wound. The fugitive couple escaped again—with dead police officers in their wake."A love story. Gunfire. Dead officers. The Bonnie and Clyde story had it all for people who couldn't get enough of their true crime stories.

While the fugitive farmer from the Midwest was a harrowing story that made the papers across the country, it could hardly compete with the Bonnie and Clyde soap opera. Soon, people got on with their lives. Investigators stopped investigating, and the lethal events of April 11, 1933, faded into history. From the fall of 1933 to the spring of 1992, when a story about the unsolved mystery appeared in the *ECM Post Review*, little if anything was written about the tragedy. Between 2008 and 2017, a handful of articles published in the *ECM Post Review*, the *Pine City Pioneer* and other news outlets sparked new interest in the case.

Public records related to the case are hard to come by. Some of the available records confirm the basics, but not much else. For example, the "Clerk's Register of Coroner's Inquest," signed in Chisago County on October 3, 1933, simply states that an inquest was held "upon the body of" Alvira Johnson, Harold Johnson, Clifford Johnson, Kenneth Johnson, Bernice Johnson, Dorothy Johnson, Lester Johnson and James Johnson. The names are hand-written in neat penmanship, but the document doesn't go into details. Many of the records, including the grand jury proceedings, have either disappeared or are stashed away on dusty shelves, unseen by modern eyes. Typically, grand jury testimony is non-public information because the law wants to protect unjustly accused suspects, witnesses and other innocent parties. "It's completely discretionary if they would want to release it," said David Schultz, the Hamline University and University of Minnesota law professor. "But almost never are grand jury deliberations released. That was true back then. That is true now." One issue: a lot of evidence that can't be used in a trial is fair game in a grand jury proceeding, Schultz said. "And back then, practically anything could be used," he said. "It could be statements, it could be hearsay, it could be who knows what? Whatever they had."

Despite the unanswered questions—or maybe because the case is shrouded in mystery—many locals are still bothered and bewildered by the dreadful fire and presumed mass murder.

10

Boozer Hank, Ghost Stories and Urban Legends

Most folks in the Twin Cities have never heard of Albin or Alvira Johnson. But more than eighty-five years after the fire, those names still resonate with many old-timers in East Central Minnesota—and even some of the younger residents.

Rush City resident Kelly Ann Hokanson was born decades after the flames and billowing smoke cast a pall over an otherwise peaceful Tuesday morning in Harris, Minnesota. She's not related to anyone on the Johnson or Lundeen side of the mystery. She doesn't have an agenda. But Hokanson grew up in the area, and she has heard all the stories. And the gregarious Kelly knows how to spin a yarn in her own right.

On a hot July morning in 2017, Hokanson offered a guided tour of the old haunts in and around Harris. Her car radio softly played classic rock tunes as she pointed out all the places of interest, from the old Krantz residence to Boozer Hank Johnson's former home. As a youngster, Hokanson often rode her horse near the old Albin Johnson place. From time to time, Boozer Hank would playfully tease her, telling her to be watchful for the mysterious big man who had allegedly slaughtered his family so many years before. "I remember sometimes Boozer Hank, when I would ride by…he would say, 'Be careful. Old Man Johnson is still out there. He's going to get you,'" Hokanson said. "It was more of a ghost story. And I was told it was five kids.…And that the husband just split."

As young Kelly understood it, Albin Johnson was similar to another legendary boogeyman in East Central Minnesota: Hinckley Harry. As that tall tale goes, an unfortunate gentleman named Harry was trying to flee the Great Hinckley Fire—a real-life disaster that killed more than four hundred people in 1894, when Albin Johnson was four years old. Harry supposedly tried to get away by jumping onto a moving train. Instead, he became trapped under the speeding locomotive, which chopped off poor Harry's legs. "And he laid there and hollered for help. And nobody came. And a wolf got a hold of him. And so he is supposedly still up there, and sometimes when the moon is full, you've got to be careful because Hinckley Harry is still out there looking for his legs," Hokanson said, completing the story in a convincing enough way that the listener almost believes it's true.

Hokanson isn't alone in equating the Johnson tragedy with a ghost story. Well into the twenty-first century, the story lives on in the minds—and perhaps vivid imaginations—of other Harris-area denizens.

A May 2017 *ECM Post Review* story about the tragedy and this book project elicited a number of comments from locals who still get goose bumps at the mention of Albin's name. Storytellers likened the fugitive farmer to a boogeyman, a merciless monster who wouldn't hesitate to intervene when young children misbehaved. One mischievous older brother went as far as to mention Albin in the same breath as mass killer Ed Gein, the notorious "Butcher of Plainfield" who terrorized the good folks of Wisconsin in the 1950s and 1960s. "Being raised with an older brother's horror stories, I heard a lot," Sandi Niemeyer wrote in a social media comment section. "I remember the story. Glenn used to alternate between telling us that Albin was going to get us [and telling] Ed Gein stories. Hadn't thought about that in years." The *Post Review* article also jogged the memory of SelenaRae Carlson, another local resident. "Remember this story well," she wrote. "Many stories in our neighborhood of this man—sleeping in barns and stealing food—for a lot of years. Even our farm."

Years later, Hokanson had another brush with the tragedy, one that was much more rooted in real life, far removed from the ghost stories of the past. At one point in her life, she worked at a nursing home in Rush City. One of the residents was an elderly Mary Johnson, wife of Big Hank Johnson, Albin's brother. Another was Ellen Scherer, one of Alvira's three sisters. Hokanson remembers Mary as a "spitfire." On one occasion, Mary Johnson was eating dinner at the old folks' home while a gentleman next to her was eyeing Mary's tater tots. He asked if she would give him one. She ignored him. He repeated his request. Same non-answer. Finally, he reached over and grabbed one. And then another. And then Mary's patience ran out. "She took the plate and flung it, like, 'Here, eat the whole damn thing why don't you?'" Hokanson said with a laugh. "We usually didn't have any trouble, but if she didn't want to do something, you realized, 'OK, we will approach you later because we are not going to push your button.'" Hokanson liked Mary. Still, "if she got frustrated, she would call us some not nice things. She never hit us or anything. But she could cut you to pieces with her tongue," Hokanson said.

Interest in the case has taken some eccentric twists. In at least one instance, the Albin Johnson case has been equated with the Lindbergh baby kidnapping—one of the most sensational crime stories of the twentieth century. A now-deceased woman who grew up near Harris contacted researcher Nan Hult a number of years back after reading a 2008 newspaper article about the Albin Johnson case. Alice Smith (not her real name) had a history of mental illness and an obsession with both the Albin Johnson mystery and the Lindbergh baby kidnapping. She claimed that Johnson had been a father figure to her. "[Mrs. Smith] was born in 1930," Hult recalled in a letter. "She says she was just two years old when she first met Albin, who was also called 'Papa' and 'Papa Indian.' So this would have been a year before the fire." Mrs. Smith referred to Albin at times as her "Daddy" and kept a voluminous collection of scrapbooks and articles about the Johnson and Lindbergh cases. Every detail of the Johnson case, in particular, was etched in her brain. "She spent years compiling [the articles] and she knew every word in each article and could rattle off without hesitation what they all meant," Hult noted. "She was so convincing." Mrs. Smith's claim that she personally knew Albin, and was like a daughter to him, should be taken with a huge dose of skepticism. Mrs. Smith was blessed with a strong imagination and a tendency to place herself at the center of historical events. Even so, her vivid stories, personal collections and musings about the Albin Johnson case speak to the intense local interest in the story.

And then there's Gerry Roll, a grandmotherly North Branch resident who devotedly pays her respects to Alvira and the children and honors their memory though she never knew the family and isn't related. Roll read about the tragedy when it was featured in a local newspaper in 2008. Since that time, she has taken it upon herself to, on a regular basis, put flowers, small toys or pinwheels on the shared grave of the eight family members.

The unsolved Harris murder mystery has also resonated with people in the Twin Cities who happened to stumble upon the story. With thinning shoulder-length hair, Al Terry looks like he just strolled out of Haight-Ashbury in search of a Grateful Dead concert. Terry lives in St. Paul, not San Francisco. But he decided on a whim to make the sixty-plus-mile drive to Rush City in June 2017 to hear a presentation at the North Chisago Historical Society about the fire. He was on his way through town a few weeks earlier when a newspaper story about the tragedy caught his eye. One thing that really struck a nerve with Terry was the unfathomable behavior of Albin's father. "I was aware from the story that the owner of the property [Emil] had evicted that family in the Great Depression....That one kid had a shoe with a hole in it. Now they are going to be homeless? It was an amazingly desperate time, it was, 'God, what is going to happen to us? Is there any future...worth seeing?' That kind of feeling." Like others, Terry was able to muster some empathy for Albin. Here was a desperate man who, along with his wife and children, had been literally put out to pasture by the patriarch of the extended family.

> *And it was a time when men were supposed to take care of their family. And someone like Albin Johnson couldn't do that. He could not do that. But he still has that reflex, that training, especially back then. And he is evicted. He failed—probably in his heart. It probably hurt so deeply. He probably felt so useless. What is going to happen to them?...Maybe* [in Albin's twisted mind] *it was more important to allow these kids to go in a way they are not going to suffer rather than face amazing suffering with a dad who really can't help them, and a mom who really can't help them. So my sense is, it is very possible that he made that decision, and it might have broke his heart. It might have broke his heart.*

Terry's also open to the theory that Albin was buried in the field next to the fire-ravaged farmhouse, perhaps deposited into that makeshift grave by his own brothers. Rumor has it that Albin's dog—a faithful pet that survived the fire—loitered in that field for days after the farmhouse burned, supposedly in grief over the loss of his master. "When a dog loves someone and that someone dies and they have access to the grave, they will lay there. They grieve like people do. And the dog will stay there, so I find that very interesting," Terry said. "It's very unfortunate that someone didn't follow up on that in a sense....The process of elimination."

Like Christine Lundeen, Alvira's mother, Terry admired President John F. Kennedy. He referenced Kennedy's assassination while trying to wrap his mind around the terrible suffering that the Lundeen and Johnson families endured after the fire. "He was like someone who was like God to me. He was better than the one they told about in catechism who was mean and set people on fire....And then he's assassinated. It took me until my forties before

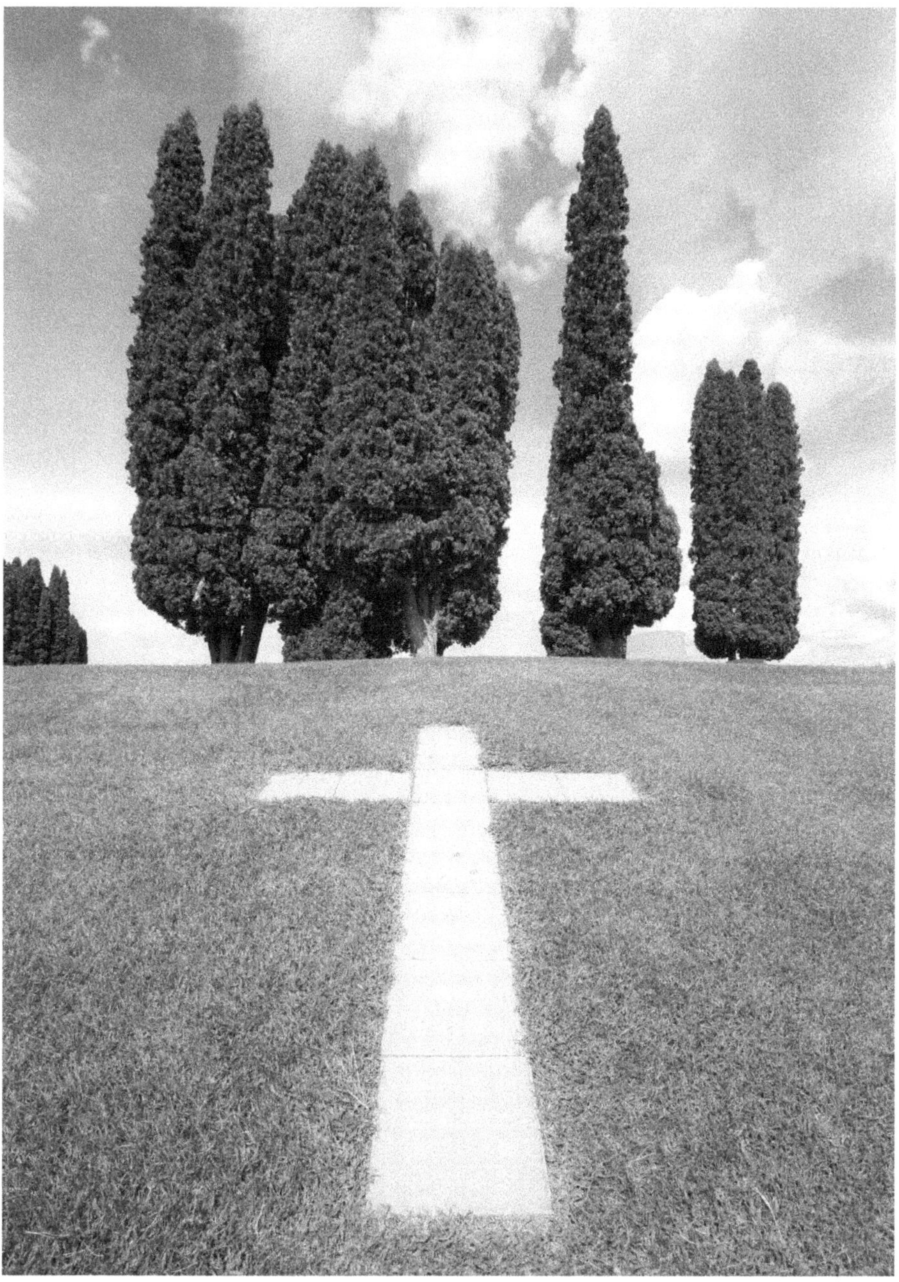

A cross is embedded in the lawn at First Lutheran Cemetery, resting place of Alvira Lundeen Johnson and her seven children. *Bill Klotz.*

I was able to let that go," Terry said. "And the human mind, automatically as I learned in college, when it has a problem it can't solve, this computer, this bio-computer, it just keeps at it. It just keeps at it. It's looking for the answer and it keeps going. And so is it any wonder that we are sitting here and we can't seem to get by this?" he said, pointing at the graves surrounding him at First Lutheran Cemetery in Rush City.

In the end, Terry believed he was meant to come face-to-face with the story of the Johnson tragedy. "And if it's any memorial to those children, and if the only thing I can do is change what's in here," he said, pointing to his heart, "that is my memorial to them."

11

Did the Brothers Do It?

Assuming Dr. Erdmann was correct in his belief that the Johnson family died of unnatural causes, Albin Johnson stands out as the obvious murder suspect. Big Albin was broke and probably despondent. He had been booted from the farm. He was indicted by a grand jury. And, of course, he bolted and was never seen again.

But not everyone's so sure that Albin Johnson was the guilty party.

Floyd Pinotti, a retired law enforcement officer who acquired the old Krantz property many years after the fire, presents an alternative theory: the idea that Albin's brothers may have had something to do with the crime. If Pinotti's theory is true, Albin was literally right under the searchers' feet the entire time they were looking hither and yon for the farmer. He theorizes that Albin himself was murdered and buried in a field adjacent to the house. A former Chisago County sheriff with thirty years of experience in law enforcement, Pinotti was going on eighty-four years old when he talked about the Albin Johnson mystery in a 2018 interview. His voice was strong over the telephone, and it was clear that Mr. Pinotti's mind was still sharp. Pinotti emphasizes that he may be reaching a bit. He doesn't claim to have all the answers. And if evidence to the contrary presents itself, he's more than willing to admit that he was wrong. "I don't have a personal preference one way or another," he said. "And it's in the Good Lord's hands."

With all of those caveats in mind, Mr. Floyd Pinotti makes his case. Speaking like a seasoned defense attorney or prosecutor, he lays out the facts one by one. Among the facts as he knows them: Albin's brothers were seen "discing" the field at 5:30 a.m. on April 11, 1933, just two hours after the fire; for days after the fire, Albin's dog ran circles around a specific location in the field, seemingly in search of his master. (Presumably, Albin had more than one dog; a different pet died in the fire.) And neighbors who knew the Johnson brothers were afraid of the boys. "I don't claim to be Sherlock Holmes, but to me, evidence is evidence," Pinotti said. "I am of the belief that Albin was buried at the property, somewhere out there where the dog was running around, and that the murders were committed by somebody who wanted the property."

Big Hank Johnson, of course, ended up with the property. Oddly enough, Pinotti himself never had a beef with Henry Johnson during all the years they lived near each other. In fact, Pinotti described Albin's brother as "a great neighbor." Pinotti added, "My children just loved him. We got along fine."

Others had a different view of Albin Johnson's kid brother. During the interview, Pinotti's mind wandered back to an incident that happened more than fifty years earlier. As Pinotti remembers it, he had just purchased a new tractor and some neighbors were gathered when he arrived home with the vehicle. The neighbors were in a relaxed mood. They were making small talk. Everything was fine. And then, Big Hank showed up. "Henry and all the other neighbors had John Deeres. And I had a gas tractor that you could hear coming in those days. When I pulled in the yard, two or three neighbors were there. And everyone was talking—'I don't know, Floyd. That thing uses a lot of gas.' And so forth....Then Henry pulled up. Those neighbors were gone in ten seconds. They were gone. It was like they vanished."

This was more than thirty years after the fire. But people in and around Harris were still jumpy and nervous when Albin's brother came calling. No doubt, Hank had a reputation, and people didn't want to take any chances. "When you are dealing with people that you perceive to be...with very little conscience, there are so many ways to retaliate on you," Pinotti said. "You can't be awake twenty-four hours a day. You've got all this exposure with your farm and your crops and your animals and your buildings. Your children. There are so many ways to make your life miserable." To the extent that people felt threatened, they couldn't rely much on law enforcement in the 1930s, Pinotti said. "This was a time when protection from the police was very rare and hard to get," he said. "You were on your own. The sheriff might come in a day or two or he might never come."

What about the manhunt that extended to Canada? Stark, Wennerberg and company had no doubt that Johnson was alive and on the run. Pinotti speculates that the popular "Albin-escaped-to-Canada" theory was fueled by a desire to take the heat off the folks in Harris farm country so people could get on with their lives and breathe a little easier. Even if they had some misgivings about the brothers, folks in town "didn't want the possibility for retaliation...either by Henry or Henry's brothers," Pinotti speculates. "It served their purpose to say he left and went up to Canada and couldn't be found."

Margie Thorp, a local resident, tells of a similar theory. Thorp's stepfather, Grant Johnson, was born in 1926 and grew up near Harris. He lived on a farm close to the Johnson place along with his brother, Filmore, and his parents, Hilda (Hessland) Johnson and Per Olaf "Ole" Johnson. "I'm thinking the blaze must have been huge, because Grant told me that they could see it from their farm. Early on that morning, Ole and little Grant (who would've been 6 years old) drove to the Albin Johnson farm...when they saw flames and smoke," Thorp recalled in an e-mail. "Grant said that the house was almost gone or just a pile of smoldering lumber."

Thorp's stepfather was stunned by what he saw early on the morning after the fire. "What stuck in his mind was that there was one of Albin's brothers plowing a field at such an early hour—the brother didn't appear to be fazed at all about the fact that the house had burned right next to him, nor did he stop to offer help," Thorp wrote.

"Grant told me that everyone in the area was afraid of the brothers, and were sure that they had killed Albin's family, set fire to the house, and then buried Albin out in the field [and] plowed him under."

Grant Johnson was related to a family named Lofgren in the Harris area, Thorp noted. One of those relatives, Augusta Lofgren, was around at the time of the fire. Fortunately for historians, she kept a diary. Mrs. Lofgren mentioned the Johnson family tragedy in an entry she recorded not long after the fact. Alluding to the rumor that the victims had been beheaded, Mrs. Lofgren put her thoughts on paper in a surprisingly unemotional narrative. She sums up the story in just seventy words:

> *The Albin Johnson house burned in the middle of the night, with Mrs. Johnson and her 7 children inside. Their heads are buried in a nearby field. Mr. Johnson's body appears to be missing and volunteers search for him. Foul play is suspected. There's an inquest in Harris. On May 1 it's reported he's been seen in Montana. In the end, the evidence isn't conclusive and no murderer is jailed.*

From there, Mrs. Lofgren abruptly segued to a far more mundane topic: the news that potatoes were selling for twenty-five cents per hundred pounds.

Lindgren, the researcher with roots in Chisago County, doesn't have much to say about the price of potatoes in 1933. But he has strong feelings about the Johnson case. Specifically, he doubts that the Johnson brothers were involved. The brothers supposedly were motivated by a desire to take over the farmland, but the facts aren't consistent with that theory, he said. "I don't think they would have burned down the house if they wanted to live there," Lindgren said. As for the story about the dog looking for his master, "if there was a dog and his house burned down, he wouldn't have anyplace else to go," Lindgren said matter-of-factly.

Henry Johnson had a place to go after he took over the property. But in the end, Big Hank suffered one setback after another. Like his brother, Henry struggled to make a go of it on the farm, Pinotti said. Drought took a toll on his crops, and his luck with livestock wasn't much better. Perhaps the property itself was cursed—insofar as it was used as farmland, at least. Those who study the history of the land will find similar stories of woe, including multiple foreclosures, Pinotti noted. Or maybe it was bad karma for whatever sins Henry Johnson may have committed as a much younger man way back when in the 1930s. "It was just one catastrophe, one bad showing after another," Pinotti said. "I love Henry because he was a good neighbor, and I hope he is resting in heaven. But I think he paid his hell right here on earth."

12

What Really Happened?

Assuming Johnson was guilty, what could possibly have driven a man to kill his wife and seven young children? Perhaps it was Albin's mindset that he could no longer support his family—and that without Albin's support, the family was better off dead. Or as the *Chisago County Press* put it in an April 20, 1933 article, "It appears like another case wherein a despondent person, discouraged with life, tries the worst to make things better." It's not unusual for despondent folks to take out their frustrations on family members. But more often than not, the killer turns the gun on himself after doing in his kinfolk. That doesn't appear to be the case with Johnson. If Johnson did make a run for it after committing the crime, where would he turn? He didn't have any money and his options were limited.

One long-shot possibility is that Johnson was a victim. Maybe he landed in hot water with some ruffians, perhaps moonshiners. He didn't have as much as a dime in his pocket. Could it be that Johnson was in debt to some unsavory characters and they took revenge on him and his family? Perhaps they disposed of Johnson's body in a remote location to throw the investigators off track. In his 1936 affidavit, Harry A. Galpin hints at the possibility of third-party involvement. If the "unexplained bones found by relatives in a distant part of the ruins" weren't Albin's, perhaps it was proof that "an accomplice might have been detected," Galpin wrote. Still, there's little proof to back up those claims.

The more conventional view is that Johnson just slipped away. Lindgren said it would have been easy in Albin Johnson's day to hide in plain sight. Law enforcement was pretty unsophisticated. Hobos, wanderers and highwaymen could easily blend into the crowd. Another factor: There was plenty of lawless activity at the time. The infamous Bonnie and Clyde were at the height of their notoriety. Perhaps law enforcement, such as it was, had more to worry about than a fugitive farmer from the heartland. "I imagine law enforcement had much bigger fish to fry," Lindgren said. "Law enforcement was completely overwhelmed by what was going on." As Lindgren put it, there were "millions of people riding the rails" in those days, desperately looking for work in the dark days of the Depression. "He could have been in Montana before they thought to look for him," he said. Moreover, the searchers wouldn't make anyone forget about Sherlock Holmes. At the time, the county sheriff's department was "an absolute joke," Lindgren said. "The only requirement to be a sheriff was to be big and tough."

"My personal opinion is Albin crossed the river into Wisconsin, and could have gotten help from his brothers, or hopped a freight or thumbed a ride to Duluth/Superior and went to parts unknown," Lindgren said. "But I guess we will never know what happened. Too bad they can't dig up the graves and do DNA and see if any parts of him are there. Never happen though."

Another longtime area resident, Greg Strom, speculates that Johnson likely would have taken his wife and seven children out of the house one at a time. There, he could have strangled each member of the family before placing their bodies back inside the house in their sleeping positions. "Obviously they had an outhouse. Maybe he woke up one kid at a time, [and said], 'Let's go out to the outhouse.'...They wouldn't have died perfectly still like they went to bed," Strom said.

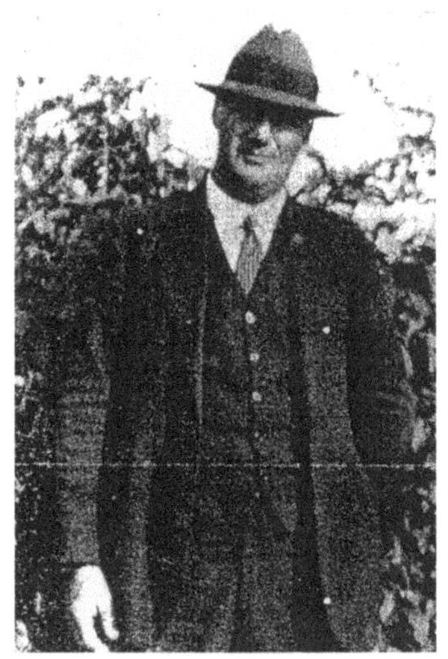

Albin Johnson, as seen in his wanted photo. Researcher Dick Lindgren speculates that Albin made his way to "parts unknown." *Pinkerton Consulting & Investigations, Inc.*

Other old-timers are inclined to believe that Albin Johnson was innocent.

Muriel Kennedy Krantz Cash isn't related to the Johnson or Lundeen families. But geographically speaking, she was as close as anyone to the fire. As a child, she lived on the farm next to the Johnson place. Her father, Ragnar Krantz, discovered the fire and then alerted the Rush City Fire Department. As of summer 2017, Muriel was still alive and well and living in the Harris–North Branch area. She still looks like the picture of health well into her eighties. She has longish, curly gray hair and a warm smile. Muriel sat down and talked about her family and the Johnson tragedy at the North Branch senior center, where she occasionally shows off her musical talents. On a warm day in July 2017, she was recruited at the last minute to play the piano in honor of a senior center worker who was leaving for a new job.

Muriel was only two years old when the fire happened, so she draws a blank when asked what she remembers about it. Her parents, of course, remembered it well. But they didn't talk much about that horrific day, Muriel said. "It was too awful," she said. "My folks were very reticent about anything. There were other things I knew that they never talked about. Not this, but in the family. Every family has some black sheep, and they didn't talk about them either." (To be clear, no one has ever suspected anyone named Krantz had anything to do with the fire.) Perhaps Muriel, like her parents, is simply too nice to entertain thoughts that an old neighbor was capable of such a monstrous deed. But she's in the group that says Albin didn't do it. So was her mother. "My mother always said that she was convinced the babies' bones were burned up completely....But she was always looking on the bright side of things. So she couldn't believe that Albin would do that."

Jeanette Johnson, Alvira's niece, tends to believe the investigators who determined that Alvira and her children were dead before the flames consumed the house. "If they had been alive, there would have been an effort to escape the fire," she reasons.

Even so, a surprising number of people were and are convinced that Albin was innocent. Alvira's sister Ellen Scherer was among those on the Lundeen side of the family who gave Albin a not guilty verdict. "Albin had a sister-in-law, Mary, who was acquainted with Ellen," Jeanette recalls. "Mary said, 'Albin loved his family, he would never do something like that.' She insisted. And she had Ellen convinced he didn't do it."

Others are skeptical of the investigators' insistence that Alvira and the children were found in their sleeping positions—a vital clue that led to the conclusion that the victims were dead before the house went up in flames. "How can they determine that they were in their sleeping positions? When the authorities got there, there was only a corner of the house left. That whole thing would have collapsed," a veteran of the Rush City Fire Department said at a North Chisago Historical Society meeting in 2017. "I've been in the fire department for thirty-one years, and I never saw that one."

And then there was Harry Galpin, Albin's devoted brother-in-law. Though Galpin's scathing affidavit in defense of Johnson fails to offer concrete evidence that Albin died in the fire, Galpin did raise some serious doubts.

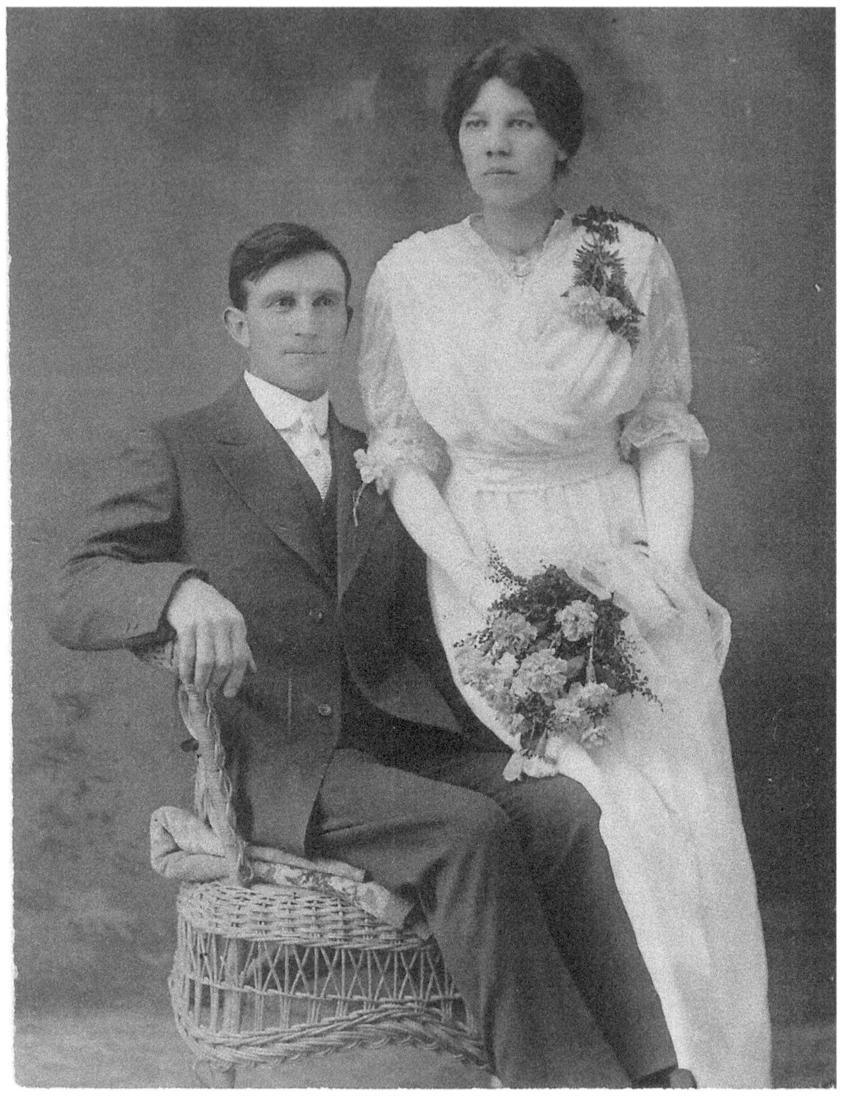

Matt Scherer and his wife, Ellen Lundeen Scherer. Ellen believed that her brother-in-law Albin Johnson was innocent. *Author's collection.*

Was there some kernel of truth in his bold assertions that the authorities and investigators were incompetent at best or corrupt at worst?

In the old days, and today in many cases, people tended to put their faith in law enforcement. County attorneys, judges, fire chiefs and medical examiners wear the white hats. They're smart. They're educated. And they live in the mansions up on the hill. Why shouldn't we take them at their word when they say something is amiss with Albin Johnson? Aren't they more credible than Johnson's ordinary friends, neighbors and kinfolk?

Now we know better. Just look at the headlines. The officer-involved shooting of Philando Castile, for example, raised serious doubts about the actions and judgment of the accused cop. Similarly, a Minneapolis neighborhood reacted with anger and disbelief when an officer shot and killed an innocent woman, forty-year-old Justine Damond, in the summer of 2017. In 2016, a judge in Seminole County, Florida, "berated and belittled" a domestic abuse victim in the courtroom, as reported in a September 2016 *Washington Post* story. The judge was dressed down by the Florida Supreme Court.

Those stories are reminders that anyone can mess up. Perhaps Harry A. Galpin realized that when he dove headfirst into the Johnson case, trying desperately to clear the name of his embattled, missing brother-in-law. Maybe Galpin was ahead of his time. Or maybe he was just blinded by his own interest in standing up for the honor of his sister's husband.

A good case can be made that the investigation into the fire and presumed murder was sloppy and incomplete, but the facts seem to indicate that Albin Johnson snapped and killed his family. The family had just been kicked off the farm, and their options were limited. It would have been quite a coincidence that the house just happened to go up in flames at precisely that moment of desperation.

As to the skepticism about the evidence that Alvira and the children were found in their sleeping positions, it should be remembered that the investigators were very clear about where the bones were found. Specifically, "remains of five of the children were found huddled near one of the doors and the mother's body lay [beside] the crib of the baby," the *Chisago County Press* reported on April 13, 1933. "Bones of another child were found in a corner of the cellar directly below where the kitchen stood."

The *Chisago County Press* offers more details. "The body of the mother, Mrs. Alvira Johnson, and the tiny babe [James]...were found in a bedroom located in the southeast corner of the house. The bodies of five of the other children were found together in an adjoining room, and the body of the oldest boy [Harold] was found in the northwest corner of the kitchen," the newspaper states.

In a 2018 interview, fire consultant Nyle Zikmund, former chief of the Spring Lake Park–Blaine–Mounds View Fire Department in suburban Minneapolis, said there's a chance that the family members could have been rendered unconscious from carbon monoxide poisoning. But it's more likely that the mother and seven children were lifeless before the fire started, as investigators concluded in 1933, he said. Fires are noisy, and it seems "extraordinary that everybody slept through it. That seems highly improbable," Zikmund said. "You wake up from heat, too. Heat wakes you up, noise wakes you up. And I have never seen scientific evidence that your olfactory senses don't work when you sleep."

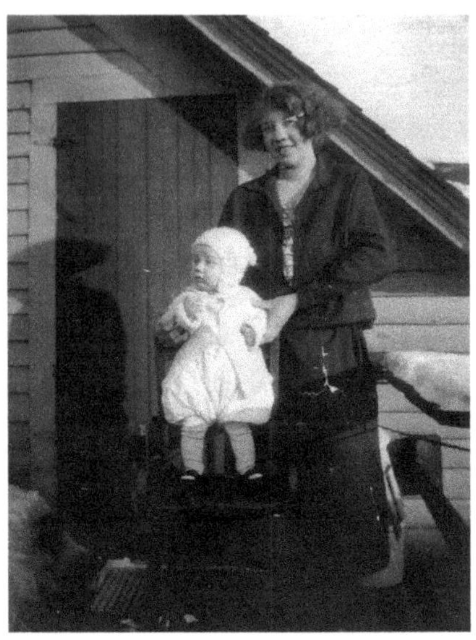
Alvira Lundeen Johnson and Harold. Alvira's body was reportedly found in a bedroom. Harold's remains were in the kitchen. *Author's collection.*

What about the potentially incriminating evidence? Wennerberg said that "two pistols and rifle" found at the scene may have been used to slay the mother and children. Pinotti saw the kerosene can in the evidence locker. More important, of course, is what they didn't find: any trace of big Albin Johnson, dead or alive.

Zikmund said investigators can rule a fire "suspicious" even if they don't have enough evidence to prove arson. In this case, "you are firmly entrenched on the suspicious pedestal because you've got way too many unexplained and abnormal behaviors," he said.

An old neighbor of the Johnsons', the late Elsie Anderson, was clearly suspicious of Albin Johnson and his motives. Mrs. Anderson, who was an adult at the time of the fire, knew Albin personally. In a 1992 interview, Mrs. Anderson said she was certain that Albin Johnson had hopped a midnight train to Canada, where he had previously worked as a logger and a farmhand. Her husband, Jack, was good friends with Fred Peterson, Alvira's brother-in-law. Elsie's daughter, the late Mae Anderson, was similarly persuaded. Unlike today, a man in 1933 could quite easily run away and assume a new identity, she reasoned in a 1992 interview with the author. Perhaps that's exactly what the fugitive farmer managed to do.

Good theories. But in the end, that's all they are—theories, assumptions, conjecture and best guesses. Betty Kollas, Alvira's niece, sums it up nicely: "There are a lot of things that could have happened. And I don't think anybody knows for sure," she said.

Epilogue

Though I've studied the Albin Johnson case top to bottom and from every conceivable angle, I'm at a loss to explain what happened to the sullen, taciturn farmer. Maybe he was consumed by the flames like the rest of his family. Perhaps he was getting blitzed on white lightning with a cigarette wedged between his calloused fingers. And then he passed out and inadvertently set the family home ablaze. Or maybe he achieved the same result by knocking over a lantern in a state of drunkenness. Anything is possible.

My best guess: Albin Johnson was clinically depressed and extremely volatile. He also had a violent streak and a tendency to overindulge in whiskey. Perhaps he was an alcoholic with no close friends, 12-step programs or mentors to turn to for help. Back in his day, people struggling with depression and the bottle didn't have many options—especially destitute farmers like Johnson. It was up to each individual to just snap out of it.

Perhaps it's also true that Johnson had narcissistic personality disorder, a malady that gives people an inflated sense of self-importance. According to the Mayo Clinic, people with narcissistic personality disorder have "a deep need for admiration and a lack of empathy for others. But behind this mask of ultra-confidence lies a fragile self-esteem that's vulnerable to the slightest criticism." Such disorders cause "problems in many areas of life, such as relationships, work, school or financial affairs," the Mayo Clinic notes on its website. "You may be generally unhappy and disappointed when you're not given the special favors or admiration you believe you deserve. Others may not enjoy being around you, and you may find your relationships unfulfilling."

Johnson was never diagnosed with depression, narcissistic personality disorder, alcoholism or any other disease of the mind as far as I can tell. But the symptoms of those illnesses—vulnerability to criticism (perhaps from father Emil Johnson), a state of general unhappiness and "problem areas" in relationships, work and finances—seem to fit Johnson neatly, much like that Scotch cap he wears in his wanted poster.

Add a diagnosis of clinical depression, narcissistic personality disorder and alcoholism to extreme poverty and hopelessness, and you have a lethal combination. Perhaps a deadly powder keg of anger, frustration and mental illness was under Johnson's skin and it all exploded during the early morning hours of April 11, 1933.

Still, the story remains an unsolved mystery.

At times, the facts surrounding the mystery bounce through my head like the silver orb in a pinball machine. At other times, they pound away like a pile driver on a construction site. Eight dead, including seven children. Only a corner of the house standing. Bodies found in sleeping positions. A missing man. Wanted posters. Reward money. An indictment. Tire tracks in the snow. An eviction notice from dear old dad. Two pistols and a rifle. A kerosene can in an evidence locker. Albin's detractors. Albin's defenders. Ghost stories. Desperation. Despair. The Great Depression. Eight lifeless, pre-cremated bodies sharing a common casket. It's an ambitious enough project just to try to sort everything out, to separate fact from fiction, truth from urban legend. Darn near impossible is the task of finding that elusive "aha!" moment, when the great mystery will be solved. Most likely, that day will never come.

Bibliography

Introduction

Johnson, Nathan. "Ghost Hunters Converge on Rush City's Grant House." *Press Publications,* March 15, 2011.
"New Hunt Made for Father in Home Fire." *St. Paul Dispatch,* April 12, 1933.
www.granthousehotelmn.com/history.
www.presspubs.com/article_080c65c4-e6eb-53a7-bcbe-31508071638e.html].
www.rushcitymn.us/.

Chapter 1

Associated Press."First Lady Dumped Into Puddle of Mud." *Bismarck Tribune,* April 13, 1933.
Cliff Bedell (Harris resident), e-mail correspondence with Nan Hult, July 13, 2007.
Dick Lindgren (former Chisago County resident), in discussion with author, 2017.
"Eight in Family Perish, Father Missing in Fire." *St. Paul Pioneer Press,* April 12, 1933.
"Family of Eight Perish in Flames." *Rush City Post,* April 14, 1933.
"Hold Jubilee as Brew On Tap in 19 States." *San Diego Evening Tribune,* April 7, 1933.
"Manhunt Underway in Manitoba." *Winnipeg Free Press,* May 5, 1933.
"Phone Rate Bill Passed in House." *Rush City Post,* April 28, 1933.
Rush City Post, April 28, 1933.
"$39,912 Allotted to This County for Use on Roads." *Rush City Post,* April 28, 1933.

Chapter 2

Associated Press. "Farmer Lodge Knocked Out by Primo Carnera." *Chicago Tribune*, February 25, 1930.

———. "New Hunt Made for Father in Home Fire." *St. Paul Dispatch*, April 12, 1933.

"At Rest." *Rush City Post Review*, March 28, 1924.

"B. Blackfelner of St. Paul Shoots Harris Cashier." *Brainerd Daily Dispatch*, April 17, 1931.

Betty Kollas (niece of Alvira Lundeen Johnson), in discussion with author, spring 2017.

boxrec.com/en/boxer/11659.

Dick Lindgren (former Chisago County resident), in discussion with author, 2017.

"Farmer Lodge Dies Following Injury Suffered in Accident." *North Branch Review*, August 20, 1941.

"Farmer Lodge Shows Tonight." *Minneapolis Star*, February 22, 1934.

"Farmer Lodge Will Come to Help Jack Dempsey's Training." *The Monroe News-Star*, August 25, 1923.

Floyd Pinotti (former Chisago County sheriff), in discussion with author, August 2018.

Harry A. Galpin, correspondence, October 1936.

Jeanette Johnson (niece of Alvira Lundeen Johnson), in discussion with author, summer 2017.

Jim Carlbom (longtime Harris area resident), in discussion with author, summer 2017.

Johnson, Brian. "Mystery Surrounding 1933 Harris Fire Still Unsolved." *ECM Post Review*, April 5, 1992.

"Johnson Missing from Scene of Harris Fire Tragedy." *Chisago County Press*, April 13, 1933.

"Last Rites Were Held at Harris For Emil Johnson." *Rush City Post*, June 11, 1948.

Mark Ruhland (relative of Hank Johnson), in discussion with author, spring 2017.

"Mother and Seven Children Burn to Death." *Chisago County Press*, April 13, 1933.

"Mr. Johnson Rites Are Held." *Rush City Post*, November 10, 1967.

Steve Hansmann (longtime Chisago County resident), in discussion with author, June 2017.

Strand, A.E. *A History of Swedish-Americans of Minnesota*. Chicago: Lewis Publishing Co., 1910.

Vreeland, W.C. "A Stride Ahead in Sports." *Brooklyn Daily Eagle*, December 21, 1923.

www.co.chisago.mn.us/DocumentCenter/View/7672/Brochure---Chisago-County-Information-Guide---2018?bidId=].

www.exploreminnesota.com/travel-ideas/swedish-heritage-itinerary/.

www.mnhs.org/mnhspress/books/emigrant-novels-book-i.

www.mnopedia.org/agricultural-depression-1920-1934.

www.thecanadianencyclopedia.ca/en/article/lumberjacks/.

Chapter 3

"A.O. Stark Service is Well Attended." *Rush City Post*, April 1961.

Associated Press. "Authorities Also Seeking Ashes of Husband in Ruins." *Bismarck Tribune*, April 12, 1933.

———. "Bones of Father Not with Family in Home's Ashes." *Duluth Herald Tribune*, April 13, 1933.

———. "Mother and 7 Children Die; Man Missing." *Tribune Republican* (Greeley, CO), April 13, 1933.

———. "Nine Perish in Flaming House." *Galveston Daily News*, April 12, 1933.

Dick Lindgren (former Chisago County resident), in discussion with author, spring 2007.

Greg Strom (longtime Harris-area resident), in discussion with author, spring 2018.

"Family of Eight Perish in Flames." *Rush City Post*, April 14, 1933.

"Farm-Labor Platform 'Death Knell' to State, G.O.P. Keynoters Say." *Minneapolis Star*, April 13, 1934.

"Father Hunted After Family Is Killed in Blaze." *Duluth Herald*, April 12, 1933.

Floyd Pinotti (former Chisago County sheriff), in discussion with author, August 2018.

"Friends from All Over the State Attend Frank W. Hanson Services." *Rush City Post*, November 6, 1942.

"Harris Centennial, 1884–1984." Harris Centennial Committee, coordinated by Irene Martinson and Inez Hanson, printed by Review Corporation.

"Harris Farmer Hunted in City." *Minneapolis Journal*, April 28, 1933.

"Harris Fire Cause Unknown, Jury Says." *St. Paul Pioneer Press*, April 15, 1933.

Jeanette Johnson (niece of Alvira Lundeen Johnson), discussion with author, July 2017.

"Johnson Missing from Scene of Harris Fire Tragedy." *Chisago County Press*, April 13, 1933.

"Mother, 7 Children Die as Farm Home Is Razed by Flames." *Minneapolis Tribune*, April 12, 1933.

Muriel Krantz Kennedy Cash (daughter of Ragnar Krantz), in discussion with author, July 2017.

"Mystery Shrouds Death of Eight in Farm Fire Tuesday Morning." *North Branch Review*, April 13, 1933.

Nan Hult (longtime North Branch resident), in discussion with author, spring 2007.

"Official Condemns Members of Mob." *Minneapolis Star*, December 8, 1933.

"The State Contest." *Manitou Messenger*, March 1913.

"300 Search for Father of 7 Dead in Fire." *Minneapolis Journal*, April 18, 1933.

United Press. "Think 8 Dead in Fire Slain." *Minneapolis Star*, April 15, 1933.

Venzell Lindholm (longtime Harris-area resident), in discussion with author, summer 2017.

www.findagrave.com/memorial/105700184/anna-wennerberg.
www.findagrave.com/memorial/search?firstname=Willard&middlename=&lastname=Krantz&birthyear=&birthyearfilter=&deathyear=2010&deathyearfilter=&location=Hawaii%2C+United+States+of+America&locationId=state_13&memorialid=&datefilter=&orderby=.

Chapter 4

Greg Strom (longtime Harris-area resident), in discussion with author, spring 2018.
"Harris Centennial, 1884–1984." Harris Centennial Committee, coordinated by Irene Martinson and Inez Hanson, printed by Review Corporation.
"Harris Farmer Hunt Launched." *Minneapolis Star*, April 27, 1933.
Hult, Nan. Unpublished narrative. "Johnson/Lundeen Tragedy of 1933."
"Police Are Continuing to Keep Sharp Lookout for Alleged Murderer." *Winnipeg Free Press*, May 8, 1933.
Sommer, Carl H. *Looking Back Over One Hundred Years in Northern Chisago County*. Rush City, MN: Rush City Post, 1958.
www.pinkerton.com/our-difference/history.

Chapter 5

Betty Kollas (niece of Alvira Lundeen Johnson), in discussion with author, spring 2017.
Christine Lundeen funeral, audio tape, January 1972.
"Death of Pioneer Resident." *Rush City Post*, November 19, 1920.
"Dies from Result of Automobile Accident." *Rush City Post*, November 15, 1929.
Hellsing family history, published in Sweden.
Jeanette Johnson (niece of Alvira Lundeen Johnson), in discussion with author, July 2017.
Johnson, Jeanette. *Hellsing-Lindstrom Family History*. Self-published, n.d.
"Last Rites for Mother, Children." *Rush City Post*, April 21, 1933.
"Pioneer Residents Called to Their Last Rest." *Rush City Post*, April 19, 1912.
Ralph Carlson (longtime area resident), in discussion with author, June 2017.
www.ecumennorthbranch.org.

Chapter 6

"Alb. Johnson Is Described." *Chisago County Press*, April 27, 1933.
"Albin Johnson Hunt Continued." *St. Paul Dispatch*, May 11, 1933.
"Albin Johnson Traced to Manitoba." *Winnipeg Evening Tribune*, May 5, 1933.
Associated Press. "Eight Victims of Fire Slain." *Evening Tribune*, Albert Lea, April 26, 1933.

———. "Seek Farmer for Slaying of His Family." *Evening Tribune*, April 27, 1933.
Dick Lindgren (former Chisago County resident), in discussion with author, 2007.
"Dr. Erdmann Is Dead at 74." *Minneapolis Star*, February 19, 1941.
"8 Dead Before Home Burned." *Minneapolis Tribune*, April 27, 1933.
"Friends View Johnson Case Without Hope." *St. Cloud Times*, May 10, 1933.
"Harris Farmer Sought in East." *Albert Lea Evening Tribune*, April 28, 1933.
Jeanette Johnson (niece of Alvira Lundeen Johnson), in discussion with author, spring 2017.
"Johnson Clue Collapses as Hunt Goes On." *St. Cloud Times*, May 6, 1933.
"Man Held by Police at Cavalier, N.D. Is Not Albin Johnson." *Winnipeg Free Press*, May 6, 1933.
"Murder Theory Is Strengthened." *Brainerd Daily Dispatch*, April 26.
"New Hunt Made for Father in Home Fire." *St. Paul Dispatch*, April 12, 1933.
"Police Believe Killer Suspect Has Ended His Life." *Winnipeg Tribune*, May 6, 1933.
"Probe Death in Family of Fire." *Ironwood Daily Globe* (Ironwood, MI), April 27, 1933.
"Royal Mounted Police Close In on Reported Hiding Place Harris, Minn. Farmer Wanted in Deaths of Family." *Brainerd Daily Dispatch*, May 5, 1933.
"Search of Fire Ruins Renewed." *Minneapolis Tribune*, April 28, 1933.
"300 Search for Father of 7 Dead in Fire." *Minneapolis Journal*, April 18, 1933.
www.tourismsaskatchewan.com/community/346/prelate#sort=relevancy.

Chapter 7

archives.fbi.gov/archives/washingtondc/press-releases/2014/fbi-adds-william-bradford-bishop-to-ten-most-wanted-fugitives-list.
Chisago County Press, May 8, 1952.
David Schultz (Hamline University law professor), in discussion with author, August 2018.
"Farmer Indicted in Death of Wife, 7 Children, in Fire at Harris, Minn." *St. Paul Dispatch*, October 30, 1933.
Greg Strom (longtime Harris-area resident), in discussion with author, spring 2018.
"Life-Long Resident Is Called by Death." *Rush City Post*, January 10, 1964.
Nan Hult (longtime North Branch resident), in discussion with author, spring 2007.
"Wife Convinced Sailstad Died in Fire at Cottage." *Minneapolis Star*, October 16, 1920.
www.fbi.gov/wanted/murders/william-bradford-bishop-jr.
www.wsj.com/articles/fbi-removes-accused-killer-of-his-family-from-ten-most-wanted-list-1530213109].

Chapter 8

Associated Press. "New Evidence Found in Fire." *St. Cloud Times*, May 4, 1933.
"Convict Galpin in Libel Attack." *Minneapolis Star*, October 15, 1939.
David Schultz (Hamline University law professor), in discussion with author, August 2018.
Harry A. Galpin, notarized affidavit, filed in Douglas County, Nebraska, October 1936.
"Harry Galpin of Harris Passes Away Suddenly." *Rush City Post*, March 1940.
"Wennerberg." Obituary, *Minneapolis Star*, January 11, 1977.
www.findagrave.com/memorial/105700184/anna-wennerberg.

Chapter 9

"Bonnie & Clyde—gangstrar, stjärnor och klantskallar?" Sveriges Radio, April 23, 2017.
David Schultz (Hamline University law professor), in discussion with author, August 2018.

Chapter 10

Al Terry (St. Paul resident), in discussion with author, June 2017.
Kelly Ann Hokanson (Rush City resident), in discussion with author, July 2017.
Nan Hult (Rush City resident), in a letter to the author, February 2016.
www.facebook.com/postreview/.

Chapter 11

Dick Lindgren (former Chisago County resident), in discussion with author, 2007.
Floyd Pinotti (former Chisago County sheriff), in discussion with author, August 2018.
Margie Thorp, e-mail correspondence with Nan Hult, May 26, 2017.

Chapter 12

Berman, Mark, and Marwa Eltagouri. "Officer Charged with Murder in Shooting Death of Unarmed Woman in Minneapolis Alley." *Washington Post*, March 20, 2018.
Betty Kollas (niece of Alvira), in discussion with author, spring 2017.
Chisago County Press, April 20, 1933.

Elsie Anderson (neighbor of the Johnsons), in discussion with author, spring 1992.

Greg Strom (longtime Harris-area resident), in discussion with author, spring 2018.

Harry A. Galpin, notarized affidavit, filed in Douglas County, Nebraska, October 1936.

Jeanette Johnson (niece of Alvira Lundeen Johnson), in discussion with author, summer 2017.

Mae Anderson (neighbor of the Johnsons), in discussion with author, spring 1992.

Muriel Krantz Kennedy Cash (daughter of Ragnar Krantz), in discussion with author, July 2017.

Nyle Zikmund (fire consultant), in discussion with author, August 2018.

Phillips, Kristine. "'Intolerable': Judge Reprimanded after She 'Berated and Belittled' Domestic-Violence Victim." *Washington Post*, September 1, 2016.

"Searching Party Hunts in Vain for Missing Man." *Chisago County Press*, April 13, 1933.

Epilogue

www.mayoclinic.org/diseases-conditions/narcissistic-personality-disorder/symptoms-causes/syc-20366662.

About the Author

Brian Johnson, a Twin Cities journalist, has more than twenty-five years of experience as a reporter. A longtime staff writer for a Minneapolis-based business publication, Johnson has won a number of awards from the Minnesota Society of Professional Journalists. The Minneapolis native graduated from the University of Minnesota in 1989 with a major in journalism and a minor in Swedish. Johnson is a grandnephew of Alvira Lundeen Johnson, who died with her seven children on April 11, 1933, in Harris, Minnesota. Johnson lives in Richfield, Minnesota, with his wife, Stephanie, and their four children.

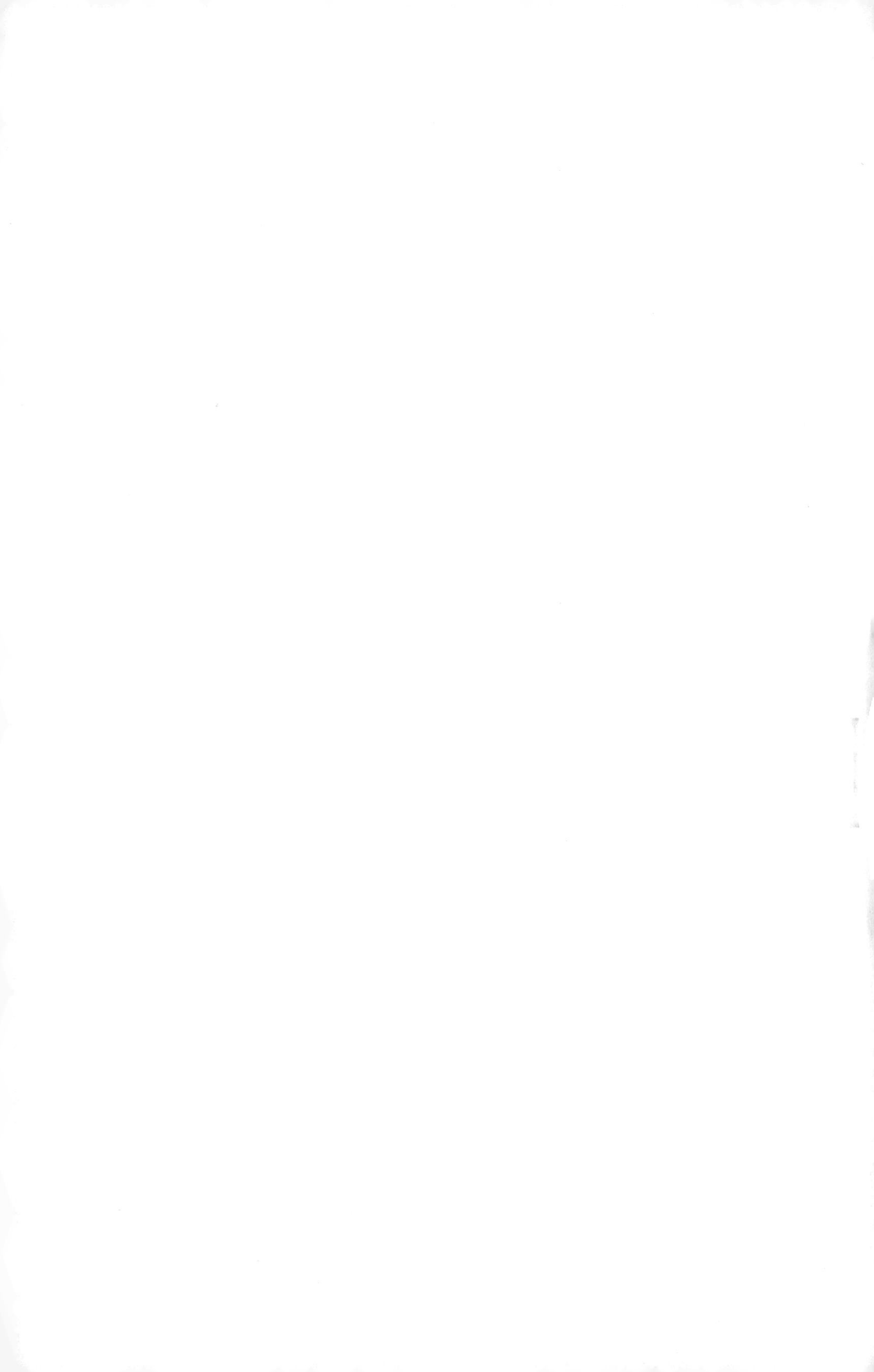

www.ingramcontent.com/pod-product-compliance
Lightning Source LLC
LaVergne TN
LVHW052342100826
845147LV00021B/1154